MW00473510

Amphibians and Reptiles of the Carolinas and Virginia

SECOND EDITION,

REVISED AND UPDATED

Amphibians

&

Jeffrey C. Beane, Alvin L. Braswell, Joseph C. Mitchell,
William M. Palmer, *and* Julian R. Harrison III

Photographs by Jack Dermid   *With contributions by*
Bernard S. Martof *and* Joseph R. Bailey

REPTILES

of the **CAROLINAS** *and* **VIRGINIA**

SECOND EDITION,
REVISED AND UPDATED

The University of North Carolina Press
CHAPEL HILL

Manteo Library
East Albemarle Regional Library
Manteo, NC 27954
252-473-2372

WITHDRAWN

This book was published with the assistance of Progress Energy.

© 2010 THE UNIVERSITY OF NORTH CAROLINA PRESS
All rights reserved

Designed by Kimberly Bryant with Giovanna de Graaff and set in Arnhem and Scala Sans by Tseng Information Systems, Inc.

Manufactured in China

The paper in this book meets the guidelines for permanence and durability of the Committee on Production Guidelines for Book Longevity of the Council on Library Resources. The University of North Carolina Press has been a member of the Green Press Initiative since 2003.

Library of Congress Cataloging-in-Publication Data
Amphibians and reptiles of the Carolinas and Virginia / Jeffrey C. Beane ... [et al.] ; photographs by Jack Dermid ; with contributions by Bernard S. Martof and Joseph R. Bailey. — 2nd ed., rev. and updated.
p. cm.
Includes bibliographical references and index.
ISBN 978-0-8078-3374-2 (cloth : alk. paper) —
ISBN 978-0-8078-7112-6 (pbk. : alk. paper)
1. Amphibians—North Carolina—Identification. 2. Amphibians—South Carolina—Identification. 3. Amphibians—Virginia—Identification. 4. Reptiles—North Carolina—Identification. 5. Reptiles—South Carolina—Identification. 6. Reptiles—Virginia—Identification.
I. Beane, Jeffrey C.
QL653.N8A4 2010
597.90975—dc22   2009039268

cloth   14 13 12 11 10   5 4 3 2 1
paper   14 13 12 11 10   5 4 3 2 1

Our dear friend and colleague Julian R. Harrison III passed away during the later editing stages of the second edition of *Amphibians and Reptiles of the Carolinas and Virginia*. The enduring contributions he made as a multidisciplinary naturalist, scientist, and educator, and the lives that he touched, are many. He achieved what most people hope to accomplish during their lives—to be loved by family and friends, to be relevant and appreciated for actions taken, and to be remembered for many positive accomplishments.

This book is dedicated to the memory of Joseph R. Bailey, Julian R. Harrison III, and Bernard S. Martof.
Their steadfast desire to pass along the natural history knowledge they had accumulated to students, colleagues, and the public was inspiring and fostered many others to pursue careers in the natural sciences. The importance of these men to science, to conservation of our natural resources, and to the betterment of humankind is tremendous. The discipline of herpetology and the natural sciences in general have been blessed by their presence and enduring legacy.

Contents

Amphibians and Reptiles of the Carolinas and Virginia

SECOND EDITION,

REVISED AND UPDATED

Introduction

Amphibians and reptiles play critical roles in natural systems, and many are highly beneficial to humans. Although these animals have long appealed to amateur naturalists as well as professional zoologists, their remarkable diversity of shapes, sizes, colors, patterns, ecologies, and life histories remain poorly known to most of the public. They constitute what has been called "hidden biodiversity" because many species are secretive and are seen rarely or only when one is actively looking for them. In the past few decades, information about amphibians and reptiles has grown tremendously. The explosive spread of urban and suburban living and outdoor recreation has evoked a resurgence of interest in the identification, natural history, behavior, and distribution of plants and animals, especially amphibians and reptiles. This book was written to acquaint persons with these abundant and varied groups of animals that live in Virginia and the Carolinas and to encourage the growth of knowledge about and understanding of these organisms and their importance. We hope it will be a useful reference not only to herpetologists and other biologists and naturalists but to all persons concerned about the environment and the quality of life in our region.

Virginia and North and South Carolina constitute a compact natural area bordered by the Appalachian Mountains to the west and northwest, the Atlantic Ocean to the northeast and southeast, and the Savannah River to the southwest. The region harbors a rich herpetofauna of some 189 species. This richness is due to several biogeographic patterns. A number of southern species have ranges extending northward through the Coastal Plain and terminating in this region. Several others have essentially northern distributions that reach their southernmost limit in the region. Numerous species (about a third of the total) have broad distributions in eastern North America, encompassing much of the mid-Atlantic region and occurring in all three states. A few species extend into the western part of the region via the Tennessee River drainage. And last, but certainly not least, some 40 salamander species have distributions centering on the Blue Ridge and Appalachian mountains. These overlapping distribution patterns yield totals of some 111 amphibian and 78 reptile species known to occur in the tristate region. About 22 species occur areawide, or nearly so, and at least that many others inhabit two-thirds or more of the area. Furthermore, the area contains 11 endemic species and several others nearly restricted to it, thus providing numerous unique elements.

A species includes a population or a group of populations whose members share many traits and are usually distinguishable from individuals of other species. Members of a species interbreed or are capable of interbreeding among themselves but are reproductively isolated from individuals of other species by a number of mechanisms (such as behavior). Some species are subdivided into geographic races, or subspecies. Such populations are morphologically or physiologically different and inhabit only a part of the total geographic range of the species. Only a few of the more conspicuous subspecies are mentioned in this book.

The species here included are those generally recognized by most herpetologists; however, for questionable or controversial species, our choices of the alternatives (often not unanimous) are used. Divergence of opinion arises mainly because some species are in different stages of evolution, and because knowledge of many populations is so fragmentary. Many species are capable of becoming divided into geographically isolated populations, each of which may accumulate genetic differences and become morphologically or physiologically distinct over time. If such a population remains geographically isolated, its taxonomic status (whether it is an unusually distinct subspecies or a full-fledged species) is often a subjective judgment. However, if the populations in question overlap on the landscape, then their taxonomic status may be easily ascertained. If, in the zone of overlap, the parental phenotypes occur frequently and hybrids only occasionally, taxonomists conclude that barriers to interbreeding exist and that the populations are separate species. On the other hand, if most individuals in the central part of the zone of overlap have some of the diagnostic features of both parental forms, the populations are best classed as the same species. The problem of recognizing species is further exacerbated because some populations have become reproductively isolated (do not interbreed) but are phenotypically similar or identical (sibling or cryptic species).

Numerous changes in the scientific names of amphibians and reptiles have occurred since the first edition of this book was published in 1980. This is largely due to the advances in molecular technology that provide insights into the genealogy of these animals. Changes in taxonomic names reflect advances in understanding genetic relationships among species or groups of species. Thus, taxonomy, the naming of species, is not a static science. Use of molecular technology has also caused scientists to rethink what constitutes a species. In short, how genetically different must a population be in order to be properly recognized as a full species? In some cases, especially in salamanders, species are recognized entirely by their genetic differences rather than by external characteristics we can see. Because of their limited mobility, salamanders are often more easily influenced by some of the factors (e.g., geographic barriers) that lead to speciation than are more mobile animals. See Table 1 for a listing of species added and name changes since publication of the first edition of this book in 1980.

Taxonomy is often controversial, and not all published taxonomic changes are immediately universally accepted in the scientific community. Examples include *Lithobates* for *Rana* (true frogs), *Anaxyrus* for *Bufo* (toads), *Plestiodon* for *Eumeces* (skinks), and *Pantherophis* for *Elaphe* (rat snakes). Because this book is designed for a general audience, we decided to remain conservative and use the older, more established names. However, we have added in parentheses in the appropriate accounts the newer names for future reference based on the 2008 checklist of North American amphibians and reptiles published by the Society for the Study of Amphibians and Reptiles. Names will continue to change as more and more groups are studied with modern techniques. In several of the accounts, we note that we anticipate that a particular species may undergo revision in the near future.

In spite of numerous attempts to standardize the common names of amphibians and reptiles, much controversy remains. An obstacle to standardization, of course, is the deep entrenchment of different names for a species in various regions of the country. In general, our common names follow those recommended in 2008 by the Society for the Study of Amphibians and Reptiles, but we have departed in a few cases where we believe our selections more appropriately describe the animal, or where we feel that retention of certain traditional, long-standing names would be more familiar and less confusing to most readers.

All species featured in this book are native to the area, with the

TABLE 1 Species Added and Scientific Name Changes since the 1980 Edition

Salamanders

Desmognathus fuscus split into:

Desmognathus conanti	Spotted Dusky Salamander
Desmognathus fuscus	Northern Dusky Salamander
Desmognathus planiceps	Virginia Dusky Salamander
Desmognathus folkertsi	Dwarf Black-bellied Salamander
Desmognathus marmoratus	Shovel-nosed Salamander (formerly *Leurognathus marmoratus*)

Desmognathus ochrophaeus split into:

Desmognathus carolinensis	Carolina Mountain Dusky Salamander
Desmognathus ochrophaeus	Allegheny Mountain Dusky Salamander
Desmognathus ocoee	Ocoee Salamander
Desmognathus orestes	Blue Ridge Dusky Salamander
Desmognathus santeetlah	Santeetlah Dusky Salamander

Eurycea bislineata split into:

Eurycea bislineata	Northern Two-lined Salamander
Eurycea cirrigera	Southern Two-lined Salamander
Eurycea wilderae	Blue Ridge Two-lined Salamander

Eurycea quadridigitata split into:

Eurycea chamberlaini	Chamberlain's Dwarf Salamander
Eurycea quadridigitata	Dwarf Salamander
Eurycea n. sp.	"Sandhills Eurycea"
Plethodon aureolus	Tellico Salamander

Plethodon glutinosus split into:

Plethodon chattahoochee	Chattahoochee Slimy Salamander
Plethodon chlorobryonis	Atlantic Coast Slimy Salamander
Plethodon cylindraceus	White-spotted Slimy Salamander
Plethodon glutinosus	Northern Slimy Salamander
Plethodon variolatus	South Carolina Slimy Salamander
Plethodon teyahalee	Southern Appalachian Salamander

Plethodon jordani split into:

Plethodon amplus	Blue Ridge Gray-cheeked Salamander
Plethodon cheoah	Cheoah Bald Salamander
Plethodon jordani	Jordan's Salamander
Plethodon meridianus	South Mountain Gray-cheeked Salamander
Plethodon metcalfi	Southern Gray-cheeked Salamander
Plethodon montanus	Northern Gray-cheeked Salamander
Plethodon shermani	Red-legged Salamander

Plethodon kentucki	Cumberland Plateau Salamander
Plethodon sherando	Big Levels Salamander
Plethodon ventralis	Southern Zigzag Salamander (formerly *P. dorsalis*)
Plethodon virginia	Shenandoah Mountain Salamander

Frogs

Bufo fowleri	Fowler's Toad (formerly *B. woodhousii fowleri*)
Pseudacris crucifer	Spring Peeper (formerly *Hyla crucifer*)
Pseudacris feriarum	Upland Chorus Frog (formerly *P. triseriata feriarum*)
Pseudacris kalmi	New Jersey Chorus Frog
Pseudacris ocularis	Little Grass Frog (formerly *Limnaoedus ocularis*)
Rana capito	Carolina Gopher Frog (formerly *R. areolata*)

Turtles

Apalone ferox	Florida Softshell (formerly *Trionyx ferox*)
Apalone spinifera	Spiny Softshell (formerly *Trionyx spiniferus*)
Kinosternon baurii	Striped Mud Turtle
Pseudemys concinna	River Cooter (formerly *Chrysemys concinna* and *C. floridana*)
Pseudemys rubriventris	Red-bellied Cooter (formerly *Chrysemys rubriventris*)
Trachemys scripta	Yellow-bellied Slider (formerly *Chrysemys scripta*)

Lizards

Hemidactylus turcicus	Mediterranean Gecko
Ophisaurus mimicus	Mimic Glass Lizard

Snakes

Heterodon platirhinos	Eastern Hognose Snake (formerly *H. platyrhinos*)
Lampropeltis getula	Eastern Kingsnake (formerly *L. getulus*)
Nerodia floridana	Florida Green Water Snake (formerly *N. cyclopion*)

exception of the Texas Horned Lizard, which has thrived in multiple established populations for several decades, and the Mediterranean Gecko, which has persisted in small colonies in all three states for several years. Small colonies of a few other nonnative species may have become established in our area as of this writing. The Brahminy Blind Snake, *Rhamphotyphlops braminus*, a tiny, parthenogenic (single-sex) species commonly transported with greenhouse plants, has been reported in parts of Richmond and Newport News, Virginia, and at least one specimen was found in a greenhouse in Wake County, North Carolina. The Chinese Softshell Turtle, *Pelodiscus sinensis*, has been reported from the Potomac River in northern Virginia. Mississippi Map Turtles, *Graptemys (pseudogeographica) kohnii*, have been found in several man-made reservoirs in central North Carolina. The Brown Anole, *Anolis* (= *Norops*) *sagrei*, has been reported from at least two localities in New Hanover County, North Carolina, and southeastern Virginia. Individual Cuban Treefrogs, *Osteopilus septentrionalis*, have been reported from several locations in the Carolinas, but no evidence of breeding has been noted. Whether these potentially invasive species will spread, maintain their populations in only small areas, or be extinguished remains to be determined. Although several other species have been introduced—in particular, such popular exotic pet species as the Green Iguana (*Iguana iguana*) and the Burmese Python (*Python molurus*) are found with increasing frequency as escaped or intentionally released individuals—none shows signs of establishing breeding populations. On the other hand, a nonnative race of a species native to our area (the Red-eared Slider, a subspecies of the native Yellow-bellied Slider, *Trachemys scripta*) has been widely introduced and is now well established in several localities. There is also evidence of intra-area and extra-area transport of native animals, chiefly the result of the use of amphibians as fish bait and reptiles as pets. More impressive changes in distributions of amphibians and reptiles are associated with agriculture, dam building, mining, coastal alterations, suburban development, and highway construction. No extinctions of our herpetofauna are known to have occurred in historical times, and we wish to keep it that way! On the other hand, numerous decreases in populations have occurred in many parts of the area due to habitat loss or other factors, and turbulent times lie ahead. This statement is as true today as it was when the first edition of this book appeared in 1980.

Recognizing that humans have drastically altered habitats and eliminated many species, Congress passed the Endangered Species Act of 1973, and subsequent reauthorizations made a variety of changes. Each state also has its own endangered species law with its own list of endangered and threatened species. There are several compelling reasons why we must prevent the extinction of species: (1) We share with other organisms a common evolutionary heritage, and we find kinship, inspiration, and beauty in many of them. (2) Human populations are large, complexly interrelated, and vulnerable to extinction. We have much to gain from studies of other imperiled populations. (3) More practically, as genetic and biochemical resources, other organisms are indispensable in biological and medical research. Clearly, the most effective protection of endangered species is provided by preservation of their natural habitats. Not only do we need more large parks and wilderness areas, but many communities would benefit immeasurably by having their own programs of habitat preservation.

To promote interest in our diminishing herpetofauna, we have listed those species requiring special protection in Table 2, which is based mainly on current state and federal listings. State listing criteria parallel, but seldom precisely mirror, federal criteria. Also, the levels of protection provided by state listing differ from federal levels and are usually less stringent. South Carolina, North Carolina, and Virginia all have legislation in place authorizing development of state listings for Endangered and Threatened species (South Carolina's In Need of Management category is the equivalent of Threatened). North Carolina has an additional Special Concern category below Threatened. Endangered and Threatened are the only official federal categories. Roughly defined, the three levels of vulnerability to extinction are indicated: Endangered species are those in imminent danger of extinction; Threatened species have reduced populations in a large portion of their ranges and are likely to become Endangered if current trends continue; and species of Special Concern have a lesser degree of endangerment and may disappear from our area or are those about which only scant information is available. Even though federal and state laws impose heavy penalties for the possession, sale, or transport of several of these species, all need the maximum protection we can provide. State and federal listings are subject to change when new information becomes available or reevaluations occur, and the necessary legal processes are followed. The most recent listings should be sought out whenever there is a need to interact with a potentially

TABLE 2 **Imperiled Species**

Species	Common Name	State Listings SC[2]	NC[3]	VA[4]	Federal Listings[1] SC	NC	VA
Salamanders							
Pseudobranchus striatus	Northern Dwarf Siren	T					
Cryptobranchus alleganiensis	Hellbender		SC				
Necturus lewisi	Neuse River Waterdog		SC				
Necturus maculosus	Common Mudpuppy		SC				
Ambystoma cingulatum	Flatwoods Salamander	E			T		
Ambystoma mabeei	Mabee's Salamander			T			
Ambystoma talpoideum	Mole Salamander		SC				
Ambystoma tigrinum	Eastern Tiger Salamander		T	E			
Aneides aeneus	Green Salamander		E				
Eurycea junaluska	Junaluska Salamander		T				
Eurycea longicauda	Long-tailed Salamander		SC				
Eurycea quadridigitata	Dwarf Salamander		SC				
Hemidactylium scutatum	Four-toed Salamander		SC				
Plethodon longicrus	Crevice Salamander[5]		SC				
Plethodon shenandoah	Shenandoah Salamander			E			E
Plethodon ventralis	Southern Zigzag Salamander		SC				
Plethodon websteri	Webster's Salamander	E					
Plethodon wehrlei	Wehrle's Salamander		T				
Plethodon welleri	Weller's Salamander		SC				
Frogs							
Hyla andersonii	Pine Barrens Treefrog	T					
Hyla gratiosa	Barking Treefrog			T			
Pseudacris brachyphona	Mountain Chorus Frog		SC				
Rana capito	Carolina Gopher Frog	E	T				
Rana heckscheri	River Frog		SC				
Crocodilians							
Alligator mississippiensis	American Alligator	T	T		all T(S/A)		
Turtles							
Sternotherus minor	Stripe-necked Musk Turtle		SC				
Clemmys guttata	Spotted Turtle	T					
Clemmys insculpta	Wood Turtle			T			
Clemmys muhlenbergii	Bog Turtle	T	T	E	all T(S/A)		

Species	Common Name	State Listings SC[2]	NC[3]	VA[4]	Federal Listings[1] SC	NC	VA
Deirochelys reticularia	Chicken Turtle			E			
Malaclemys terrapin	Diamondback Terrapin		SC				
Gopherus polyphemus	Gopher Tortoise	E					
Caretta caretta	Loggerhead Sea Turtle	T	T	T	T	T	T
Chelonia mydas	Green Sea Turtle	T	T	T	T	T	T
Eretmochelys imbricata	Hawksbill Sea Turtle	E	E	E	E	E	E
Lepidochelys kempii	Kemp's Ridley Sea Turtle	E	E	E	E	E	E
Dermochelys coriacea	Leatherback Sea Turtle	E	E	E	E	E	E
Apalone spinifera spinifera	Eastern Spiny Softshell		SC				

Lizards

Species	Common Name	SC	NC	VA	SC	NC	VA
Eumeces anthracinus	Coal Skink	T					
Ophisaurus mimicus	Mimic Glass Lizard		SC				
Ophisaurus ventralis	Eastern Glass Lizard			T			

Snakes

Species	Common Name	SC	NC	VA	SC	NC	VA
Heterodon simus	Southern Hognose Snake	T	SC				
Nerodia sipedon williamengelsi	Carolina Water Snake		SC				
Opheodrys vernalis	Smooth Green Snake		SC				
Pituophis melanoleucus	Pine Snake		SC				
Micrurus fulvius	Eastern Coral Snake		E				
Crotalus adamanteus	Eastern Diamondback Rattlesnake		E				
Crotalus horridus	Timber Rattlesnake		SC	E[6]			
Sistrurus miliarius	Pigmy Rattlesnake		SC				

[1]Federal official listings are Endangered (E) and Threatened (T) and can carry the designation S/A (similarity of appearance) for a species or population that is listed because it looks very similar to a listed species or population.

[2]South Carolina's official state listing designations are Endangered (E) and In Need of Management (T) (the equivalent of Threatened).

[3]North Carolina's official state listing designations are Endangered (E), Threatened (T), and Special Concern (SC).

[4]Virginia's official state listing designations are Endangered (E) and Threatened (T).

[5]Considered by some to be a variant of *Plethodon yonahlossee*, Yonahlossee Salamander.

[6]Coastal Plain population only.

listed species. State and federal wildlife agencies will have the most up-to-date listings.

Inasmuch as this book may inspire inquisitiveness in its readers, a few words about conserving, collecting, and keeping amphibians and reptiles are provided. No longer can the forests, fields, lakes, and streams and their inhabitants be exploited without regard for the future and for the interests of others. Many laws have been passed in response to the growing awareness of the importance of natural resources: (1) Several decades ago, song and game birds received federal protection, and this now extends to all but a few introduced avian species. (2) The taking of fur and game mammals is closely regulated by state laws. (3) Recently, many endangered plants and animals received federal and state protection. As yet, these laws embrace relatively few species of amphibians and reptiles, but many more will probably be included in the future.

Legal restrictions also apply to the collecting of all organisms in state and federal parks (including the Blue Ridge Parkway), seashores, wildlife refuges, management areas, and (without permission) all private lands. Every naturalist—the serious student as well as the casual amateur—must respect the laws pertaining to the flora and fauna in each state; however, it is not enough to observe merely the letter of the law. If collecting and field investigation (even when done within the law) result in alteration of the environment, damage to segments of the biota can result. As nearly as possible, leave the habitat as you found it. Put back into its original location each rock, log, board, etc., that you moved. They offer shelter not only to amphibians and reptiles but to a host of other important organisms. Do not tear apart decomposing logs with abandon or strip them of bark or moss, especially in areas used by the public. Keep in mind that many species are active at night and can then be collected with no disturbance to the habitat. In fact, under suitable atmospheric conditions, some species of amphibians can be seriously exploited at night. Limit your collecting to the minimum; take only as many animals as you can care for properly, and give every consideration for the future of the population. Avoid killing animals on the roads and encourage others to refrain from such senseless destruction of our fauna.

If an animal is to be kept as a pet, have proper quarters ready in advance, and remember that maintaining a pet is a long-term responsibility, not a game. Provide proper cover, as most amphibians and reptiles are very secretive. Compared with birds and mammals, amphibians and reptiles require only occasional meals (usually once or twice a week is adequate), but many need warmth and moisture. Generally, the common, readily available species are the easiest to maintain and observe. If you find that you cannot provide regular care for live animals, attempt to find someone else who can assume their care. Short-term captives that have not been exposed to disease may be returned to their original habitat—but check state regulations; some states prohibit release after an individual has been in captivity for more than several weeks. Generally, release of captive pets to the wild is not beneficial to the pets or wild populations. Help to establish more rigid controls over commerce of the pet trade and hobbyists. Keep abreast of local, state, and federal issues related to natural resources and take an informed and thoughtful position on those issues.

The species accounts in this book are fairly standardized. First, the size range of adults is given: snout to vent for frogs, carapace length (straight line from anterior to posterior tip of carapace) for turtles, and total length (tip of snout to tip of tail) for all other groups. Metric measurements are used, and English approximations are given in parentheses. A brief description of the species follows, often with comments on features distinguishing the species from similar ones. The geographic range and the habitat are briefly described in the next paragraph. This is followed by information on food habits, life history, and general biology. A color photograph of a typical individual of each species and a map showing its distribution within the area are provided. The text, photograph, and map for most species are all on one page, but for some species with unusual variation or interesting behavior, the accounts are longer and encompass two or three pages and usually two or three photographs. Because the sea turtles are strictly marine and come ashore only to lay eggs, maps of their ranges are omitted.

To facilitate finding information in the book, orders and families appear in phylogenetic sequence, and genera and species are in alphabetical order. Each class, order, and suborder is briefly described in an introductory account that includes comments on the local composition of that taxonomic group.

To identify an amphibian or reptile, compare it with the photographs provided, select the photos that most closely match the animal you have, and then carefully read the descriptions provided at the beginning of each account. Remember that just as persons differ, so do members of other species of animals. Therefore, do not be disturbed if your individual differs slightly from the one pictured. Keep in mind, too, that individuals of many species often become darker (more melanistic) as they grow older. Also, be sure to check the maps, because distribution is often an aid to identification, but remember that some geographic ranges are imperfectly known and that some are changing.

To confirm an identification or to report an extension of geographic range, write or visit the primary repository for herpetological information in your state: North Carolina State Museum of Natural Sciences, Research Laboratory, MSC #1626, Raleigh, NC 27699-1626 (www .naturalsciences.org); Charleston Museum, 360 Meeting St., Charleston, SC 29403 (www.charlestonmuseum.org); the North Carolina Herpetological Society (www.ncherps.org); or the Virginia Herpetological Society (http://fwie.fw.vt.edu/VHS/index.html). Provide the following information for all specimens: date of collection, precise location (including coordinates when possible), brief description of habitat, name of collector, and comments on behavior or activity.

Carefully study the photographs of the venomous snakes so that you can quickly identify these species in the field. Do not handle venomous snakes. Most bites occur when people handle them, attempt to kill or capture them, or pick them up when they believe the snake is dead. Be certain that you are adequately equipped and informed when working with any reptile or amphibian. Exercise caution in capturing turtles and large nonvenomous snakes; many will bite in self-defense.

ACKNOWLEDGMENTS

We thank the many individuals who made the preparation of this book so pleasant and enjoyable. A special debt of gratitude is due the many students who through the years encouraged this project not only with their general interests in amphibians and reptiles but also with their observations and collection of specimens. Much information was derived from specimens in the North Carolina State Museum of Natural Sciences, the Charleston Museum, and the Savannah Science Museum. We are grateful to these institutions for making their holdings available. While space does not permit the listing of the many persons who have rendered assistance in various ways, we are especially in debt to Stanley L. Alford, Rudolph G. Arndt, the late Joseph M. Bauman, Stephen H. Bennett, William S. Birkhead, Charles R. Blem, Cindy Bogan, the late Katherine M. and James W. Braswell, Samuel Braswell, the late Elmer E. Brown, Richard C. Bruce, E. D. Bruner, J. H. Carter III, Cheryl K. Cheshire, David G. Cooper, John E. Cooper, J. Edward Corey III, Deb Creech, Mark Danaher, Robert A. Davis, Michael E. Dorcas, Floyd L. Downs, Navar Elliott, David A. Etnier, C. Keith Farmer, John T. Finnegan, Rufus W. Gaul Jr., Matthew Godfrey, Gabrielle J. Graeter, the late M. Frank Groves, Jeffrey G. Hall, Stephen Hall, Dennis W. Herman, Richard Highton, Chris Hobson, Erich L. Hoffman, Richard L. Hoffman, Stephanie J. Horton, Thomas E. Howard Jr., C. R. Hoysa, W. Jeffrey Humphries, Robert B. Julian, the late Carl Kauffeld, Kenneth Kraeuter, Trip Lamb, David S. Lee, Harry E. LeGrand Jr., Daniel F. Lockwood, Marcia Lyons, Jonathan P. Micancin, the late Sherman A. Minton Jr., Richard R. Montanucci, Philip Moran, Peter Morin, Robert H. Mount, Nora A. Murdock, Robert Palmatier, James F. Parnell, Russell Peithman, Jesse P. Perry III, Harvey Pough, L. Todd Pusser, Tiffany A. Pusser, William H. Redmond, Jerald H. Reynolds, Valeria K. Rice, the late Joseph K. Rose III, William H. Rowland Jr., Albert E. Sanders, David T. Sawyer, Tammy B. Sawyer, Harvey Scarborough, Vincent P. Schneider, Frank J. Schwartz, David M. Sever, Nathan A. Shepard, Michael A. Sisson, Natalia L. Smith, S. Daniel Smith, F. F. Snelson Jr., Ann B. Somers, Mark Spinks, Wayne C. Starnes, David L. Stephan, Kim C. Stone, Sherrie Storm, Charles D. Sullivan, Thomas J. Thorp, Stephen G. Tilley, Franklin J. Toby Jr., Joseph A. Travis, the Tregembo family (George, June, Paul, and Robert), R. Wayne Van Devender, Kendrick Weeks, Henry M. Wilbur, Lori A. Williams, Gerald K. Williamson, David Wojnowski, Gary Woodyard, and Joseph S. Zawadowski. Lisa C. Yow assisted greatly in the preparation of the distribution maps. Joe Mitchell thanks the many herpetologists who have worked in Virginia over the past century for their work on the state's herpetofauna, as well as his family for their support of his fieldwork and time devoted to the study of amphibians and reptiles. Many people associated with the Savannah River Ecology Laboratory (J. Whitfield Gibbons and associates) have contributed, and continue to contribute, much to the knowledge of South Carolina and regional herpetology. Numerous members of the North Carolina Herpetological Society and faculty and students of Davidson College have made important contributions. Professional staff in many state, federal, and private organizations have increased their knowledge and awareness of the regional herpetofauna and contribute important information toward its documentation and conservation. Notable organizations include the state heritage programs, wildlife and fisheries agencies, environmental monitoring agencies, U.S. and state forestry units, state and national parks, The Nature Conservancy, environmental education centers and museums, and land trusts, just to mention a few.

The Area

Virginia and the Carolinas are highly diverse in topography, vegetation, and climate. As a consequence of these factors and the states' latitudinal position, the area has a remarkably rich herpetofauna. Its nearly 324,000 square kilometers (200,000 square miles) vary from sandy barrier islands with maritime forests along the coast and low, pine-covered ridges separated by floodplain forests or swamps farther inland to relatively rugged mountains dominated primarily by northern hardwoods along the northwestern margin. Between these extremes is a broad, gently rolling plateau forested with oaks and pines.

Much of the area is cultivated or urbanized, and almost none of the remainder harbors forests such as those of precolonial times. No part of the area has escaped alteration, and most forests have been logged several times. Inland waters comprise only about 5 percent of the surface area. Rivers, streams, and swamps are relatively numerous, but there are few natural lakes such as Lake Drummond in southeastern Virginia and the bay lakes in southeastern North Carolina. In recent decades, however, several large impoundments have been constructed for hydroelectric power, recreation, and flood control; typical of these are the Santee-Cooper lakes in South Carolina and Kerr Reservoir along the border between North Carolina and Virginia.

The ranges of amphibians and reptiles are dynamic, not static, and are influenced by many complex factors. As environments change, ranges may also change. A knowledge of the more salient features of the area's physiography, vegetation, and climate is prerequisite to an understanding of the habitats and ranges occupied by particular species.

PHYSIOGRAPHY

Portions of five physiographic provinces occur in the area (see fig. 1). As used here, the term "Mountains" refers collectively to the higher elevations of the Appalachian Plateau, Blue Ridge, and Valley and Ridge provinces. N. M. Fenneman's *Physiography of Eastern United*

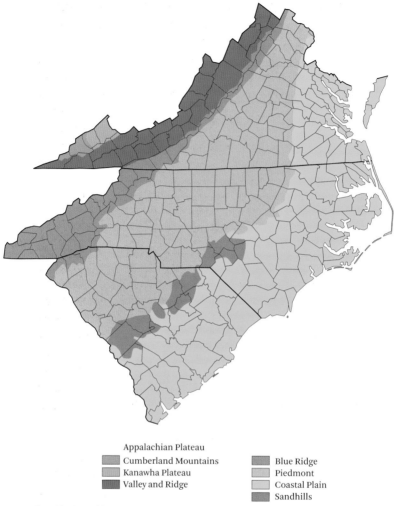

Appalachian Plateau

Cumberland Mountains   Blue Ridge
Kanawha Plateau   Piedmont
Valley and Ridge   Coastal Plain
    Sandhills

Fig. 1 Physiographic Provinces

States, published in 1938, is a useful reference for additional details concerning each province, as is the introduction to *Virginia's Endangered Species*, by S. L. Woodward and R. L. Hoffman, published in 1991. The *Appalachian Plateau*, in extreme western Virginia, is the smallest of the physiographic regions and consists of two different sections. The northeastern *Kanawha section* is a plateau with moderate to strong relief, whereas the southwestern *Cumberland Mountain section* is characterized by numerous deep valleys separated by steep, narrow ridges. Elevations average 600 m (1,970 ft.) throughout the province but range from 900 to 1,200 m (2,950 to 3,940 ft.) in parts of the Cumberland Mountains. Many streams, tributaries of the Cumberland, Kentucky, and Big Sandy rivers, flow westward toward the Mississippi River. Extensive beds of coal underlie this province.

The *Valley and Ridge province* extends southwest to northeast along the northwestern margin of Virginia. In the southwest it abuts the Appalachian Plateau, and on the east, the northernmost part of the Blue Ridge province. The Valley and Ridge province is a composite of six river valleys, including the New River and the Shenandoah River, the largest in the region. The valleys are interspersed between elongate, parallel ridges 300 to 600 m (980 to 1,970 ft.) in elevation. Massanutten Mountain, one of the highest, separates most of the Shenandoah Valley into two parts. Overall, the province is a highly folded, hilly region with a width of about 120 km (77 mi.) in the north and about 80 km (50 mi.) in the south. Sandstone, shale, and limestone are widespread, and caves are numerous.

The *Blue Ridge province* is only several kilometers wide in the area north of Roanoke, Virginia. To the southwest, especially in North Carolina, it reaches a width of more than 80 km (50 mi.) and is characterized by northeast-southwest-oriented ridges connected by a ladderlike series of cross ridges. Relief is rugged in places, accentuated by high peaks and steep, narrow gorges. The highest peaks occur in North Carolina, where 82 exceed 1,525 m (5,000 ft.) and 43 exceed 1,830 m (6,000 ft.). Mount Mitchell, in the Black Mountain range, reaches 2,037 m (6,681 ft.) and is the highest peak in the eastern United States. South Carolina is the least mountainous of the three states; only its northwestern corner lies within the Blue Ridge province, where elevations vary from 300 to 600 m (980 to 1,970 ft.). The highest peak, Sassafras Mountain, reaches 1,083 m (3,552 ft.). In Virginia, most elevations are between 450 and 1,000 m (1,475 and

3,280 ft.); exceptions include the two highest peaks, Mount Rogers at 1,743 m (5,717 ft.) and Whitetop Mountain at 1,682 m (5,517 ft.). The Blue Ridge province is composed chiefly of granites and gneisses and metamorphosed sedimentary rocks, principally siltstones, sandstones, and conglomerates. The eastern continental divide generally extends along the eastern crest of the southern portion of the Blue Ridge province; streams, such as the New River, originating on the western flank, flow into the Gulf of Mexico via the Ohio and Mississippi rivers or the Tennessee River drainage. Those originating on the eastern flank flow into the Atlantic Ocean. In Virginia, this pattern changes from the James River northward, with the continental divide passing westward through Blacksburg and northeastward in the Appalachian Mountains along the Virginia–West Virginia state line. The eastern slope of the Allegheny Mountains and the Shenandoah Valley drain northward to the Potomac River and the Chesapeake Bay.

The *Piedmont province* is a gently rolling plateau about 65 km (40 mi.) wide in northern Virginia, 225 km (140 mi.) wide at the Virginia–North Carolina border, and about 175 km (110 mi.) wide in the Carolinas. Elevations decrease from 305–460 m (1,000–1,510 ft.) in the northwest to 60–150 m (200–490 ft.) in the southeast. It is the largest of the five provinces, including about 40 percent of the total area. The Piedmont–Blue Ridge boundary is marked in places by steep escarpments through which rivers have cut deep, narrow gorges, especially in the Carolinas. The Piedmont is delineated from the Coastal Plain by the Fall Line, a narrow zone marked chiefly by rapids formed in streams as they leave the crystalline bedrock of the Piedmont and enter the Coastal Plain. Typically, there is no distinguishable change in relief at points of contact between the two provinces. While most of the Piedmont consists of well-rounded hills and elongate, rolling ridges, it is punctuated by higher monadnocks, such as the Uwharrie and King's mountain ranges in North Carolina, Parson's Mountain in South Carolina, and Willis Mountain in Virginia. The Piedmont province is geologically older than the Blue Ridge province and is also underlain by complex igneous and metamorphic rocks.

The *Atlantic Coastal Plain province* includes about 10 percent of the land area in the continental United States, and approximately 38 percent of our area. Three major divisions are relatively distinct. The westernmost division, or *upper Coastal Plain*, extends eastward from the Piedmont to the 84-m (275-ft.) contour. It has a rolling topography and is generally well drained. A subdivision, the *Sandhills*, characterized by deep, sandy soils, rolling topography, and the highest elevations in the Coastal Plain, borders the Fall Line in South Carolina and in south central North Carolina. The *middle Coastal Plain* lies between the 84-m (275-ft.) and 29-m (95-ft.) contours; it has a gently rolling topography and is also generally well drained. The easternmost division, or *lower Coastal Plain*, extends from the 29-m (95-ft.) contour to sea level. This is a region of low relief, poor drainage, and numerous swamps. In southeastern Virginia, eastern North Carolina, and southeastern South Carolina, it is bordered by sandy outer banks or barrier islands. The Coastal Plain is underlain by geologically young, unconsolidated sedimentary formations that thicken from west to east. These are covered in most places by a relatively thin layer of sands and clays.

VEGETATION

The fully developed or climax vegetation of the Appalachian Plateau is highly diverse mixed hardwood forest on most slopes, with patches of spruce on the higher peaks and knobs. Bogs are numerous in valleys at the higher elevations, and open stands of pine exist in areas of sandy soil on the lower ridges.

Forests dominated by Chestnut Oak are prevalent throughout the Valley and Ridge province, especially on ridges and the upper slopes. Pines often occupy the drier sites, and mixed hardwoods occur in moist ravines and on the steeper northern slopes. Where not farmed, valley floors harbor forests dominated by oaks and Tulip Poplar, with Flowering Dogwood as an important understory species. Eastern Redcedar is often present in limestone areas.

The Blue Ridge has the most diverse vegetation of the provinces in the area. Several types of deciduous hardwood forests are present, and stands of spruce or fir occur on many peaks that exceed 1,370 m (4,500 ft.). A highly diverse cove hardwoods forest occupies sheltered valleys, protected lower slopes, and relatively open north- or east-facing slopes, at elevations of 460 m to 1,370 m (1,475 to 4,500 ft.). Most stands contain 25 to 30 species and are similar to the mixed hardwood forests of the Appalachian Plateau and portions of the Valley and Ridge province. Hemlock forests occur along streams at lower elevations and in some moist ravines at higher elevations. These usually are not pure stands but contain various hardwoods such as Sugar Maple and Fraser's Magnolia. Dense thickets of rhododendron at the borders of streams are also characteristic. The more exposed slopes up to elevations of 1,370 to 1,525 m (4,500 to 5,000 ft.) are generally occupied by oak forests. A shrub layer is usually present but varies in thickness and in species composition; common understory species include Flowering Dogwood and Sourwood. Until the 1920s, American Chestnut was abundant and codominant with the oaks throughout the Appalachians. However, the chestnut succumbed to the attack of an introduced blight fungus from China and, except for occasional saplings growing from old stumps, is now almost extinct.

On most dry, open ridges and on steep, open, south- or southwest-facing slopes, pines replace hardwoods, especially at lower elevations. From southwestern Virginia throughout much of western North Carolina, stands of spruce or fir are found at elevations above 1,370 m (4,500 ft.). When destroyed by fire, wind damage, or cutting, the spruce-fir association is replaced by an early successional stage dominated by American Mountain-ash, Yellow Birch, and Fire Cherry. On steep south-facing gaps between stands of fir or spruce, beech orchards may be present, while at lower elevations on steep, exposed slopes with relatively rugged relief, extensive heath balds dominated by rhododendron and Mountain Laurel often develop. Throughout the southern portion of this province, open grassy areas, or grass balds, usually surrounded by forests, are common on southern, southwestern, and western exposures at elevations of 1,585 to 1,770 m (5,200 to 5,800 ft.).

Except on the wettest and driest sites, most upland portions of the Piedmont province are covered by oak-hickory forests with understories of Flowering Dogwood, Sourwood, and other species. On the drier sites, pines prevailed previously, and along rivers and streams were narrow corridors of bottomland hardwoods with thick understories of cane. On some moist, steep, north-facing slopes, usually near streams, mixed hardwood forests similar to those of the Blue Ridge province occurred. Stands of hemlock were also present in some areas, usually on bluffs along streams, but this tree is declining due to an introduced Asian insect, the Hemlock Woolly Adelgid. Until the colonial period, at least in South Carolina, open, prairielike tracts dominated by stands of cane 1.5 to 9 m (5 to 30 ft.) tall occurred on some Piedmont ridges. Although remnants of most of these early associations persist, much of the province is under cultivation or occupied by stands of second- or third-growth oaks and pines.

Mountain ridge with coniferous forest

Heath bald with rhododendron

Pine forests of various kinds dominate most of the Coastal Plain province, particularly on the more elevated, sandier, and drier sites. If forest fires are infrequent, hardwoods replace pines on the moister sites. In the southern and more coastal portions of the province, a mixed evergreen-hardwood association is the climax forest type. Farther north and in more inland areas, deciduous hardwoods replace evergreens as the dominant species. On dry, sandy sites with rolling topography, relatively open stands of Longleaf Pine and Turkey Oak may occur, although most of this habitat type has been lost since colonial times. The herb layer of these stands is dominated by clumps of Wiregrass interspersed between bare patches of white or pale yellow sand. On gently sloping ridges, or on extensive, poorly drained areas with little relief, pine savannas or flatwoods often develop. These are characterized by Loblolly or Longleaf pines with a continuous, highly diverse herb layer of grasses and other flowering plants. Wiregrass sometimes grows on the better-drained sites.

Scattered throughout upland pine savannas or flatwoods are numerous evergreen shrub bogs (also called bays or pocosins). These are often circular or elliptical, vary in size, and are commonly bordered by dense thickets of evergreen shrubs such as Sweetbay Magnolia, Loblolly Bay, Gallberry, and Sweet Pepper-bush. Pond Pine is frequently present. All of these plants may occur in the interior as well as at the margins of the bogs. Evergreen shrub bogs are usually wet throughout most of the year, but surface waters may be hidden by mats of sphagnum moss or other plants. Cypress or gum ponds are also characteristic of upland pine savannas or flatwoods. In these ponds the herb and shrub layers are sparse or absent; cypresses dominate if water is present throughout most of the year, while gums dominate if water availability is seasonal.

Bottomland or swamp forests occur along rivers and the larger streams. The floodplains of blackwater rivers or streams, and flat areas associated with upland drainage channels, are dominated by cypress or gum. Cypress is prevalent in areas where the soil rarely dries. The floodplains of major rivers originating in the Piedmont or Mountain regions usually have hardwood forests. Distinctive maritime forests characterized by evergreen oaks, hollies, Red Bay, and pines occur on the Outer Banks and barrier islands. Cabbage Palmetto is a conspicuous dominant on most barrier islands in South Carolina. Behind the primary dunes on many barrier islands, a nearly impenetrable thicket

Sandhills vegetation

Pine flatwoods

Evergreen shrub bog (pocosin)

of Yaupon, Wax-myrtle, Coastal Red-cedar, and greenbrier borders the forest.

Additional details concerning most of the habitats discussed above may be found in the following references:

Barry, John M. 1980. *Natural Vegetation of South Carolina*. University of South Carolina Press, Columbia.

Clay, James W., Douglas M. Orr Jr., and Alfred W. Stuart, eds. 1975. *North Carolina Atlas: Portrait of a Changing Southern State*. University of North Carolina Press, Chapel Hill.

A Description of the Geology of Virginia, <http://csmres.jmu .edu/geollab/vageol/vahist/PhysProv.html> (accessed 17 November 2008).

Fenneman, Nevin M. 1938. *Physiography of the Eastern United States*. McGraw-Hill, New York.

Nelson, John B. 1986. *The Natural Communities of South Carolina: Initial Classification and Description*. South Carolina Wildlife and Marine Resources Department, Columbia.

Schafale, Michael P., and Alan S. Weakley. 1990. *Classification of the Natural Communities of North Carolina. Third Approximation*.

Cypress-gum association

Maritime forest

North Carolina Department of Environment, Health, and Natural Resources, Division of Parks and Recreation, Natural Heritage Program, Raleigh.

Woodward, Susan L., and Richard L. Hoffman. 1991. The Nature of Virginia, in *Virginia's Endangered Species,* coordinated by K. Terwilliger. McDonald and Woodward Publishing Company, Blacksburg, Va.

CLIMATE

The area's climate is variable. Cold winters and relatively mild summers occur in the north and in the Mountains, while mild winters and hot summers are the norm in the south and on the coast. Coastal areas are subject to the modifying influence of the adjacent Atlantic Ocean, while the more variable patterns of the Mountains depend on local differences in topography and relief. Piedmont localities typically have cold to moderately cold winters and hot summers, with the severity of winter weather increasing toward the north and the west. Phenological events such as the initiation of breeding activity in salamanders generally occur earlier at lower elevations and more southern latitudes than at higher elevations and more northern latitudes.

In the Coastal Plain, annual precipitation averages 100 to 120 cm (39 to 47 in.) in the north and 115 to 135 cm (45 to 53 in.) in the south; the highest amounts occur in North Carolina. Precipitation is more variable in the Mountains, averaging 80 to 125 cm (31 to 49 in.) in the north and 100 to 215 cm (39 to 85 in.) in the south. The highest averages occur in the southern Mountains, particularly in the Great Smokies, the Nantahalas, and the Blue Ridge escarpment along the western border of the Carolinas. Rainfall in these areas may exceed 255 cm (100 in.) per year. Precipitation in the Piedmont is generally intermediate between that of the Mountains and the Coastal Plain. For the area as a whole, the most rain occurs in midsummer; the least occurs in the fall, and moderately high amounts occur in winter and spring. Piedmont summers are generally dry. Winter snowfall is frequent in Mountain and Piedmont areas, but mountain peaks lack snowcaps during summer months.

Average temperatures are variable in the Mountains but relatively uniform in the Piedmont and Coastal Plain. In the Coastal Plain, January averages range from 2.2–5.6°C (36–42°F) in the north to 8.9–10.0°C (48–50°F) in the south. For Mountain localities, the range is from 0–

3.3°C (32–38°F) in the north to 1.1–5.6°C (34–42°F) in the south. In the Coastal Plain, July averages range from 25.6°C (78°F) in the north to 26.7°C (80°F) or above in the south. For the Mountains, the range is between 20 and 23.3°C (68 and 74°F) at most localities. Piedmont averages closely approximate those of the Coastal Plain.

Coastal areas receive the potentially devastating effects of hurricanes from June to November, most often in September. The coast of the eastern United States endures about five hurricanes in an average year. Hurricanes affect the entire area encompassed by Virginia and the Carolinas, including the Mountains.

A Brief History of Herpetology in the Carolinas and Virginia

Three factors have combined to make the herpetological history of our area a rich one. The first is historical: the seaboard of Virginia and the Carolinas was explored and colonized early—as early as more northern states and earlier than states to the south. The second factor is biological: the amount of the fauna itself, numerous in species and individuals, is three times the size of that of New England and twice that to be found from Delaware northward. Our fauna includes such spectacular elements as alligators, sea turtles, rattlesnakes, amphiumas, and vociferous and colorful frogs, all of which were duly reported in early travelers' accounts, sometimes faithfully and sometimes with the colorful embellishment of folklore.

The final factor is one of chance: the human element. Three men, all of whom chose Charleston as a home base, figured most prominently: Mark Catesby in the pre-Linnaean era; Dr. Alexander Garden, a collaborator of Linnaeus in the 1760s; and Dr. John Holbrook, who wrote the first comprehensive North American herpetology nearly a century later.

Our herpetofaunal history began almost with the earliest settlements. The colonization of our states coincided with a burgeoning interest in natural history in Europe, where individuals and local societies were avidly collecting seeds, plants, and animal curiosities for their gardens and private collections. Under the influence of such naturalists as John Ray in seventeenth-century England and Carolus Linnaeus of Sweden in the eighteenth century, travelers and local correspondents were encouraged to send natural history specimens back for cultivation and study. Growing interests in medicine, pharmacology, and new food sources were important stimuli, as was curiosity about natural science per se. Thomas Harriot and John White were the first to report on natural history in the New World on Roanoke Island (*A Briefe and True Report of the New Found Land of Virginia* [1588]). Harriott noted consumption of turtles and their eggs by American Indians, and White included illustrations of several reptiles in his famous watercolor paintings. Captain John Smith in his *Map of Virginia*, published originally in 1612, made the first observations of reptiles in Virginia, including American Indians' use of rattlesnake rattles, live Green Snakes worn in earlobes, and consumption of sea turtles.

Subsequently, our herpetological history divides into two major phases. The first embraced the colonial period and the first years of the nineteenth century, a time of exploration and discovery, the results of which were reported by European naturalists in the European press and at scientific meetings. This period could be divided into pre- and post-Linnaean, but such a division would recognize only the formality of binomial nomenclature following Linnaeus. Neither the direction nor the quality of natural history research changed appreciably.

After a virtual hiatus of a quarter-century, during which national intellectual and scientific seeds were germinating, the second phase began when the rising school of American naturalists took over from their overseas colleagues. They announced their findings at meetings and in the journals of newly founded American societies. This phase is divided loosely into two periods. The first extends from about 1830 to World War II; during this period scientific research emphasized anatomy, development, and discovering new species of amphibians and, with the advent of Darwin's influence, evolutionary relationships. After World War II the emphasis shifted to ecological and behavioral approaches, as well as evaluation of genetic relatedness and phylogenies with molecular tools.

The early colonial period was largely a recording of travelers' tales of encounters with the local animals interspersed with the legends and beliefs surrounding them. In 1669, John Lederer (*The Discoveries of John Lederer, in Three Several Marches from Virginia to the West of Carolina* [London, 1672]) told the tale of a rattlesnake in eastern Virginia that, by the power of its stare, forced a squirrel to descend a tree to be eaten. Robert Beverly (*The History and Present State of Virginia* [London, 1705]) commented on several herpetological subjects: the noise made by croaking frogs, the size and "roaring" of bullfrogs, the rattlesnake (which he never saw in life), and several other snakes. He almost correctly asserted that the venomous snakes bear living young whereas other snakes lay eggs. The striking of a gun butt by the sharp tail of a horn snake with such force as to engage the butt, and the charming of birds and squirrels by snakes are among his anecdotes. He also told of a rattlesnake encountered in the process of swallowing a squirrel, the other end of which was taken by a dog that pulled until the snake gave up. The dog ate the squirrel, and the raconteur ate the snake, "which was dainty food"—perhaps the first account of what was to become an exotic hors d'oeuvre for the adventurous gourmet two and a half centuries later.

The first really significant account of the herpetofauna of our area is that of John Lawson, surveyor, naturalist, and outdoorsman, whose *A New Voyage to Carolina* (London, 1709; reprinted as the *History of Carolina* [London, 1714]) was based on his travels and experiences a few years earlier. His list "Insects of Carolina" included the alligator, about twenty kinds of snakes, three lizards, a tortoise and terrapin, "many sorts" of frogs, and the "Eel-snake" (*Amphiuma*). At that time the terms "insect" and "reptile" referred to crawling creatures rather than to particular animal groups, and Lawson simply reversed our present usage. Beyond his list he devoted about ten pages to a discussion, species by species, of their biology, dangers, uses, and so forth, much of it accurately reporting personal observations. He gave a dramatic account of a large alligator in late March that started bellowing about 9:00 in the evening from its hibernating burrow directly beneath his cabin on the river bank. An Indian companion returned shortly after the episode to explain the source of the noise to the badly frightened Lawson. He added a description of the eggs and commented on the use of alligator flesh for food and for its medicinal properties.

Lawson's care as an observer is revealed in his descriptions of the rattle and teeth of the rattlesnake, but his credulity is also unmasked by his acceptance of its power to charm and the curative virtues of its cast skin, rattle, and gall. His comments on the remainder of the fauna are an equally amusing mixture of fact and fiction. Lawson had planned an extensive study of the natural history of Carolina, but he was murdered by Indians in 1711.

For nearly a century other travelers recorded notes and anecdotes of a similar nature, often cribbed from Lawson and seldom adding significant new information. Likewise, many serious naturalists, including John and William Bartram, Alexander Wilson, and John LeConte, passed through, but their herpetological observations in our area were very minor.

The most outstanding figure of the pre-Linnaean era was Mark Catesby, whose monumental *The Natural History of Carolina, Florida and the Bahama Islands* (2 vols.) (London, 1731–43) was in its time the most elaborate and ambitious study in natural history of the New

World. The first volume was devoted principally to birds, with descriptions and full-page paintings with natural backgrounds where the flora is shown. The second volume includes a fair representation of the herpetofauna similarly treated, as well as mammals and some marine fishes and invertebrates.

Catesby, at age twenty-seven, was interested but not greatly experienced in natural history. He went to Virginia in 1712 to stay for a time with his sister and her husband, Dr. William Cocke, who had migrated a few years earlier to become a prosperous pioneer of Virginia medicine and an early figure in the social and political life of Williamsburg. Here Catesby met William Byrd, who hosted him and encouraged his naturalistic pursuits and introduced him to other prominent persons who were to aid his endeavors. Catesby collected botanical specimens for Samuel Dale and perhaps other gardeners and nurserymen in England. He spent most of his early stay in the tidewater regions of Virginia, but he did ascend the James River to the mountains and in 1714 made a voyage to the West Indies. Little is known of his activities during the next five years, prior to his return to England in the autumn of 1719; but lasting impressions had been made, he had gained New World experience, and he had made some bird drawings.

After considerable planning and uncertainty, Catesby returned to America in 1722 with the blessings of the Royal Society and the financial backing of some of its members who hoped for returns in the form of botanical materials and natural history curiosities. On his part, Catesby's aim was to study the flora and fauna of the country and prepare illustrations for a book on its natural history. He settled in Charles Town (now Charleston) with the inducement of £20 a year authorized by the governor of Carolina. He remained there over two years interspersed with several expeditions into the Piedmont.

The Florida component of Catesby's work is vague, and it seems he himself did not engage in fieldwork south of the Savannah area. But regional names at that time lacked the precision they merit today. In 1725 he visited the Bahamas, where he concentrated on marine life, but he did draw and describe three lizards and one snake of West Indian species. All the other reptiles and amphibians depicted are species found in the general vicinity of Charleston: 1 salamander, 4 frogs, 3 lizards, 18 snakes, and 4 sea turtles. The majority of these can be identified as species recognized today, but a few snakes are possibly composite, such as the "Black Viper," which combines characteristics of the Cottonmouth and the Eastern Hognose Snake, neither of which are otherwise certainly represented. Many of Catesby's accounts were later incorporated into the binomial descriptions of species by Linnaeus and post-Linnaean authors, who, because our scientific nomenclature dates from the tenth edition of *Systema Naturae* (1758), are credited with the original descriptions rather than Catesby.

In the tenth edition of *Systema Naturae*, only two species of reptiles and amphibians were included that are thought to have been based on specimens from our area—the Five-lined Skink and the Eastern Box Turtle. When, in 1766, the next (twelfth) edition appeared, it included thirteen additional new species from Carolina. The difference was due to the efforts of Dr. Alexander Garden, a young Scotch physician who immigrated to Charleston in 1752. Dr. Garden practiced medicine for almost thirty years until he was forced to return to Britain in 1782 as a consequence of his loyalist sympathies. However, during his years in the American colony his great love and avocation was natural history, and he sent to his scientific colleagues in Europe quantities of seed and seedlings and, to Linnaeus, many preserved specimens of fishes, amphibians, and reptiles. The only species named credited to Garden himself is *Amphiuma means*, which acquired scientific status in 1821 with the posthumous publication of his correspondence with Linneaus. His specimen and letters describing it had arrived too late for inclusion in the *Systema Naturae* of 1766. Most of his few and scattered publications dealt with medical subjects, but his extensive correspondence with British and other European naturalists, as well as the drawings and collections he sent, did much to promote knowledge of our flora and fauna. Garden was responsible for first notices of such genuine novelties as the Greater Siren, the Two-toed Amphiuma, and the Florida Softshell.

Between 1789 and 1803, an additional nineteen now-recognized species from our area, predominantly frogs and turtles, were added by no fewer than eleven authors from a variety of sources. This flurry marks the end of European domination of American natural sciences in general and of Carolina herpetology in particular. Other than the *Amphiuma* mentioned above, only one final species, the Green Anole, was described by a European—Friedrich S. Voigt, in 1832.

In 1822 Dr. John Edwards Holbrook settled in Charleston to practice medicine, just as had Alexander Garden seventy years earlier. Holbrook was a South Carolinian by birth but had been raised in New

England and educated at Brown University and the University of Pennsylvania, as well as in Edinburgh and Paris. In Paris a stay at the Jardin des Plantes brought him in contact with leading figures in biology who were influential in determining his later direction in natural history. Shortly after his arrival in Charleston, he became first professor of anatomy (1824) at the Medical College of the State of South Carolina, and about the same time he began work on *North American Herpetology*, destined to become a classic. The first edition appeared in four volumes (1836–40); but Holbrook was not pleased with it, and nearly all the copies were recalled to be replaced with a second edition in 1842, in five volumes. Only four complete sets of the first edition are known to exist, and a good set of the second edition would bring several thousand dollars today. The five volumes total nearly 700 pages, with 147 individual accounts, most of which are illustrated in color from living specimens. A facsimile reprint of Holbrook's five volumes was published by the Society for the Study of Amphibians and Reptiles in 1976 (see <http://www.ssarherps.org/pages/publications.php>).

Although the work was intended to cover the herpetofauna of the whole country, then made up of twenty-five states, Holbrook's Charleston focus and the cultural heritage of the eastern seaboard assured that the fauna of the Carolinas and Virginia figured prominently. Nearly 60 percent of Holbrook's species were recognized to occur in these states, which today have just over a third of the recognized species of the present coterminous United States.

Perhaps not surprisingly, the herpetology of the area was due for a period of relative inactivity following Holbrook's masterpiece. Between then and the advent of the twentieth century another half-dozen species were described from our fauna by Edward D. Cope and others, and the late nineteenth century saw the completion of Cope's great monographs, the *Batrachia* (1889) and the posthumous *Reptilia* (1900), which updated the field for the country as a whole but had no local emphasis.

In late 1880 the Brimley brothers immigrated from England to North Carolina to devote their long lives to North Carolina's natural history. H. H. Brimley headed the new state museum in Raleigh, concentrating his efforts in ornithology and conservation. C. S. Brimley became an entomologist with the Department of Agriculture, to which field he contributed extensively and well. In addition he became the state's leading authority on herpetology. Between 1895 and 1939 he published forty-seven papers dealing with many aspects of the state's herpetofauna, including identification and description, ecology, behavior, and seasonal distribution. In the period 1939–43, he published a series of summary papers in *Carolina Tips*, and these were later brought together in a pamphlet, published by Carolina Biological Supply Company, that comprises the first herpetology of the state.

E. Burnham Chamberlain played a similar role in South Carolina herpetology. A naturalist from boyhood, he joined the staff of the Charleston Museum in 1924 as curator of vertebrate zoology and began an active research career. He is coauthor of *South Carolina Bird Life* and is best known as an ornithologist, but he contributed several valuable papers on amphibians and reptiles of South Carolina. He is largely responsible for the nucleus of the herpetological collection of the museum. Perhaps his greatest contribution is the inspiration he provided two generations ago to a sizable group of Charleston youngsters who went on to make their mark as leading herpetologists. As emeritus curator of vertebrate zoology, Chamberlain continued to be active until his death in 1986.

Another contemporary who was to have a major impact on the regional herpetofauna was Emmett Reid Dunn, a young Virginian who, early in his career, specialized in salamanders. No other area of the world comes close to having the diversity and abundance of salamanders that our area boasts. The publication of Dunn's *Salamanders of the Family Plethodontidae* in 1926 proved a tremendous stimulus to others to investigate this group. His book brought together a difficult literature in a convenient and thorough manner and called attention to the problems and the importance of the southern Appalachians as a center of abundance and diversification for this largest salamander family. The advent of the automobile and the highway network must be recognized as a major factor in exploiting the salamander fauna of these mountains, since many of the distributions are so remarkably restricted that only in the present generation have several species been discovered. Dunn described several species of salamanders, including the Yonahlossee Salamander from the mountains of western North Carolina in 1917. He contributed extensively to the herpetology of his home state.

As the contributors to our herpetological history changed, so also did the character of their contributions. The colonial period was naturally one of exploration and inventory. What sort of animals inhabited this new land? What were they good for or what hazards did they present? Descriptions and folklore embellished the lists, and gradually more accurate observations on natural history and behavior were added.

In the nineteenth century the emphasis passed to anatomy and evolutionary relationships, in keeping with prevailing trends in biology. These classical disciplines are still in vogue and being pursued actively with the addition of new genetic, biochemical, and microscopic techniques. The mainstream of present-day investigations is in population biology: community structure, species interactions, the role of reptiles and amphibians in the ecosystem, and conservation and management. Environmental preservation has become a public concern; we hope in time to save some of our more precarious species from the fate of the Passenger Pigeon, Carolina Parakeet, and Ivory-billed Woodpecker.

Space does not permit mention of the many competent persons now engaged in the study of our herpetofauna. Many of the universities in the region, as well as the Charleston Museum, North Carolina State Museum, the Highlands (North Carolina) and Mountain Lake (Virginia) biological stations, the Savannah River Ecology Laboratory, and a host of other institutions, governmental agencies, and public and private organizations, are actively sponsoring education, conservation, and research in the field. The Virginia Herpetological Society (formed in 1958) is a group of amateur and professional herpetologists contributing to our knowledge of the distribution and habits of the regional fauna. The North Carolina Herpetological Society (formed in 1978) has developed into an active group of amateurs and professionals who have contributed much to our understanding of that state's amphibians and reptiles. The North Carolina chapter of Partners in Amphibian and Reptile Conservation (formed in 2004) is one of the most active chapters of that partnership in the country.

For further information on herpetological history, see (1) *Dr. Alexander Garden of Charles Town*, by David Berkeley and Dorothy Smith Berkeley (University of North Carolina Press, 1969); (2) *Mark Catesby, the Colonial Audubon*, by George Frederick Frick and Raymond Phineas Stearns (University of Illinois Press, 1961); (3) *North American Herpetology*, by John Edwards Holbrook (provided with a biographical sketch and technical updating, the second edition was reprinted by the Society for the Study of Amphibians and Reptiles in 1976); (4) *The Reptiles of Virginia*, by Joseph C. Mitchell (Smithsonian Institution Press, 1994); and (5) *Reptiles of North Carolina*, by William M. Palmer and Alvin L. Braswell (University of North Carolina Press, 1995).

List of Amphibians and Reptiles of the Carolinas and Virginia

Class Amphibia

Order Caudata

Sirenidae
  Pseudobranchus striatus — Northern Dwarf Siren
  Siren intermedia — Lesser Siren
  Siren lacertina — Greater Siren
Cryptobranchidae
  Cryptobranchus alleganiensis — Hellbender
Proteidae
  Necturus lewisi — Neuse River Waterdog
  Necturus maculosus — Common Mudpuppy
  Necturus punctatus — Dwarf Waterdog
Amphiumidae
  Amphiuma means — Two-toed Amphiuma
Ambystomatidae
  Ambystoma cingulatum — Flatwoods Salamander
  Ambystoma jeffersonianum — Jefferson Salamander
  Ambystoma mabeei — Mabee's Salamander
  Ambystoma maculatum — Spotted Salamander
  Ambystoma opacum — Marbled Salamander
  Ambystoma talpoideum — Mole Salamander
  Ambystoma tigrinum — Eastern Tiger Salamander
Salamandridae
  Notophthalmus viridescens — Eastern Newt
Plethodontidae
  Aneides aeneus — Green Salamander
  Desmognathus aeneus — Seepage Salamander
  Desmognathus auriculatus — Southern Dusky Salamander
  Desmognathus carolinensis — Carolina Mountain Dusky Salamander
  Desmognathus conanti — Spotted Dusky Salamander
  Desmognathus folkertsi — Dwarf Black-bellied Salamander
  Desmognathus fuscus — Northern Dusky Salamander
  Desmognathus imitator — Imitator Salamander
  Desmognathus marmoratus — Shovel-nosed Salamander
  Desmognathus monticola — Seal Salamander
  Desmognathus ochrophaeus — Allegheny Mountain Dusky Salamander
  Desmognathus ocoee — Ocoee Salamander
  Desmognathus orestes — Blue Ridge Dusky Salamander

Desmognathus planiceps	Virginia Dusky Salamander
Desmognathus quadramaculatus	Black-bellied Salamander
Desmognathus santeetlah	Santeetlah Dusky Salamander
Desmognathus welteri	Black Mountain Salamander
Desmognathus wrighti	Pigmy Salamander
Eurycea bislineata	Northern Two-lined Salamander
Eurycea chamberlaini	Chamberlain's Dwarf Salamander
Eurycea cirrigera	Southern Two-lined Salamander
Eurycea guttolineata	Three-lined Salamander
Eurycea junaluska	Junaluska Salamander
Eurycea longicauda	Longtail Salamander
Eurycea lucifuga	Cave Salamander
Eurycea quadridigitata	Dwarf Salamander
Eurycea wilderae	Blue Ridge Two-lined Salamander
Eurycea n. sp.	"Sandhills Eurycea"
Gyrinophilus porphyriticus	Spring Salamander
Hemidactylium scutatum	Four-toed Salamander
Plethodon amplus	Blue Ridge Gray-cheeked Salamander
Plethodon aureolus	Tellico Salamander
Plethodon chattahoochee	Chattahoochee Slimy Salamander
Plethodon cheoah	Cheoah Bald Salamander
Plethodon chlorobryonis	Atlantic Coast Slimy Salamander
Plethodon cinereus	Red-backed Salamander
Plethodon cylindraceus	White-spotted Slimy Salamander
Plethodon glutinosus	Northern Slimy Salamander
Plethodon hoffmani	Valley and Ridge Salamander
Plethodon hubrichti	Peaks of Otter Salamander
Plethodon jordani	Jordan's Salamander
Plethodon kentucki	Cumberland Plateau Salamander
Plethodon meridianus	South Mountain Gray-cheeked Salamander
Plethodon metcalfi	Southern Gray-cheeked Salamander
Plethodon montanus	Northern Gray-cheeked Salamander
Plethodon punctatus	Cow Knob Salamander
Plethodon richmondi	Southern Ravine Salamander
Plethodon serratus	Southern Red-backed Salamander
Plethodon shenandoah	Shenandoah Salamander
Plethodon sherando	Big Levels Salamander
Plethodon shermani	Red-legged Salamander
Plethodon teyahalee	Southern Appalachian Salamander
Plethodon variolatus	South Carolina Slimy Salamander
Plethodon ventralis	Southern Zigzag Salamander
Plethodon virginia	Shenandoah Mountain Salamander
Plethodon websteri	Webster's Salamander
Plethodon wehrlei	Wehrle's Salamander
Plethodon welleri	Weller's Salamander

Plethodon yonahlossee	Yonahlossee Salamander
Pseudotriton montanus	Mud Salamander
Pseudotriton ruber	Red Salamander
Stereochilus marginatus	Many-lined Salamander

Order Anura

Pelobatidae	
Scaphiopus holbrookii	Eastern Spadefoot
Bufonidae	
Bufo americanus	American Toad
Bufo fowleri	Fowler's Toad
Bufo quercicus	Oak Toad
Bufo terrestris	Southern Toad
Hylidae	
Acris crepitans	Northern Cricket Frog
Acris gryllus	Southern Cricket Frog
Hyla andersonii	Pine Barrens Treefrog
Hyla avivoca	Bird-voiced Treefrog
Hyla chrysoscelis	Cope's Gray Treefrog
Hyla cinerea	Green Treefrog
Hyla femoralis	Pine Woods Treefrog
Hyla gratiosa	Barking Treefrog
Hyla squirella	Squirrel Treefrog
Hyla versicolor	Gray Treefrog
Pseudacris brachyphona	Mountain Chorus Frog
Pseudacris brimleyi	Brimley's Chorus Frog
Pseudacris crucifer	Spring Peeper
Pseudacris feriarum	Upland Chorus Frog
Pseudacris kalmi	New Jersey Chorus Frog
Pseudacris nigrita	Southern Chorus Frog
Pseudacris ocularis	Little Grass Frog
Pseudacris ornata	Ornate Chorus Frog
Microhylidae	
Gastrophryne carolinensis	Eastern Narrow-mouthed Toad
Ranidae	
Rana capito	Carolina Gopher Frog
Rana catesbeiana	American Bullfrog
Rana clamitans	Green Frog
Rana grylio	Pig Frog
Rana heckscheri	River Frog
Rana palustris	Pickerel Frog
Rana sphenocephala	Southern Leopard Frog
Rana sylvatica	Wood Frog
Rana virgatipes	Carpenter Frog

Class Reptilia

Order Crocodilia

Alligatoridae
  Alligator mississippiensis   American Alligator

Order Testudines

Chelydridae
  Chelydra serpentina   Common Snapping Turtle
Kinosternidae
  Kinosternon baurii   Striped Mud Turtle
  Kinosternon subrubrum   Eastern Mud Turtle
  Sternotherus minor   Stripe-necked Musk Turtle
  Sternotherus odoratus   Eastern Musk Turtle
Emydidae
  Chrysemys picta   Painted Turtle
  Clemmys guttata   Spotted Turtle
  Clemmys insculpta   Wood Turtle
  Clemmys muhlenbergii   Bog Turtle
  Deirochelys reticularia   Chicken Turtle
  Graptemys geographica   Common Map Turtle
  Malaclemys terrapin   Diamondback Terrapin
  Pseudemys concinna   River Cooter
  Pseudemys rubriventris   Red-bellied Cooter
  Terrapene carolina   Eastern Box Turtle
  Trachemys scripta   Yellow-bellied Slider
Testudinidae
  Gopherus polyphemus   Gopher Tortoise
Dermochelyidae
  Dermochelys coriacea   Leatherback Sea Turtle
Cheloniidae
  Caretta caretta   Loggerhead Sea Turtle
  Chelonia mydas   Green Sea Turtle
  Eretmochelys imbricata   Hawksbill Sea Turtle
  Lepidochelys kempii   Kemp's Ridley Sea Turtle
Trionychidae
  Apalone ferox   Florida Softshell
  Apalone spinifera   Spiny Softshell

Order Squamata

SUBORDER SAURIA
Gekkonidae
  Hemidactylus turcicus   Mediterranean Gecko

Polychrotidae
  Anolis carolinensis   Green Anole
Phrynosomatidae
  Phrynosoma cornutum   Texas Horned Lizard
  Sceloporus undulatus   Eastern Fence Lizard
Teiidae
  Cnemidophorus sexlineatus   Six-lined Racerunner
Scincidae
  Eumeces anthracinus   Coal Skink
  Eumeces fasciatus   Five-lined Skink
  Eumeces inexpectatus   Southeastern Five-lined Skink
  Eumeces laticeps   Broad-headed Skink
  Scincella lateralis   Ground Skink
Anguidae
  Ophisaurus attenuatus   Slender Glass Lizard
  Ophisaurus compressus   Island Glass Lizard
  Ophisaurus mimicus   Mimic Glass Lizard
  Ophisaurus ventralis   Eastern Glass Lizard

SUBORDER SERPENTES
Colubridae
  Carphophis amoenus   Eastern Worm Snake
  Cemophora coccinea   Scarlet Snake
  Coluber constrictor   Black Racer
  Diadophis punctatus   Ring-necked Snake
  Elaphe guttata   Corn Snake
  Elaphe obsoleta   Rat Snake
  Farancia abacura   Mud Snake
  Farancia erytrogramma   Rainbow Snake
  Heterodon platirhinos   Eastern Hognose Snake
  Heterodon simus   Southern Hognose Snake
  Lampropeltis calligaster   Mole Kingsnake
  Lampropeltis getula   Eastern Kingsnake
  Lampropeltis triangulum   Eastern Milk Snake, Scarlet Kingsnake
  Masticophis flagellum   Eastern Coachwhip
  Nerodia erythrogaster   Red-bellied Water Snake
  Nerodia fasciata   Banded Water Snake
  Nerodia floridana   Florida Green Water Snake
  Nerodia sipedon   Northern Water Snake
  Nerodia taxispilota   Brown Water Snake
  Opheodrys aestivus   Rough Green Snake
  Opheodrys vernalis   Smooth Green Snake
  Pituophis melanoleucus   Pine Snake
  Regina rigida   Glossy Crayfish Snake
  Regina septemvittata   Queen Snake

Rhadinaea flavilata	Pine Woods Snake
Seminatrix pygaea	Black Swamp Snake
Storeria dekayi	Brown Snake
Storeria occipitomaculata	Red-bellied Snake
Tantilla coronata	Southeastern Crowned Snake
Thamnophis sauritus	Eastern Ribbon Snake
Thamnophis sirtalis	Eastern Garter Snake
Virginia striatula	Rough Earth Snake
Virginia valeriae	Smooth Earth Snake

Elapidae

Micrurus fulvius	Eastern Coral Snake

Viperidae

Agkistrodon contortrix	Copperhead
Agkistrodon piscivorus	Cottonmouth
Crotalus adamanteus	Eastern Diamondback Rattlesnake
Crotalus horridus	Timber Rattlesnake
Sistrurus miliarius	Pigmy Rattlesnake

Class Amphibia

Amphibians evolved from fishes during late Devonian times and became the first land-dwelling vertebrates. They are the ancestors of the tetrapods (reptiles, birds, and mammals). Even today, amphibians are of major biological importance. They avidly eat insects and are eagerly eaten by many organisms. Their hardiness, small size, low metabolism, and simple environmental needs make them excellent experimental animals. In this capacity, they have contributed greatly to our understanding of embryology, anatomy, tissue regeneration, physiology, and behavior.

Most amphibians are four-legged; have smooth, moist skin; lack scales; lay shell-less eggs in freshwater or in moist places on land; have an aquatic larval stage, and are quasi-terrestrial as adults. The term *Amphibia* is derived from Greek and means "both life" in reference to the aquatic larval phase and the adult terrestrial phase.

Our amphibian fauna is totally indigenous. No species has been successfully introduced, but some species have been transported within the area, primarily for use as fish bait. The major disruptive factor has been human technological encroachment. Many species have been eliminated from parts of our area, but by the same token, a few have prospered, for example, with the extensive construction of farm ponds. The living amphibians of our area include the two major orders: Caudata (salamanders) and Anura (frogs and toads).

Order Caudata

SALAMANDERS

Salamanders have elongated bodies. The long trunk and tail provide the chief locomotor thrust when swimming, whereas their small limbs permit crawling on land. Some of our salamanders (Seepage and Pigmy) become adults when only about 44 mm (1.75 in.) long, but others (Hellbender, Greater Siren, Two-toed Amphiuma) approach or exceed 1 m (3 ft.). Most are about 125 mm (5 in.) long.

Almost all salamanders are voiceless. The chemical messengers (pheromones) secreted by hedonic glands evoke courtship and spawning behavior. Fertilization is external in the Hellbender and probably in the sirens. In all other species it is internal. The male deposits a packet of sperm on a gelatinous stalk (spermatophore), and the female clasps the packet with her cloaca.

Salamanders typically undergo an aquatic larval stage lasting from a few days to several years, but in some plethodontids the larval stage is completed within the egg membrane prior to hatching. The larval stage ends with metamorphosis—a series of definitive changes in structure and life cycle. Adults of some species retain numerous larval features; the sirens and mudpuppies are classical examples. Some species are totally aquatic, many live in moist places on land but go to the water to breed, and others are completely terrestrial. Surprisingly few are arboreal; possibly that habitat is too hostile even though western North Carolina has unusually high rainfall. Much variation in respiration also occurs. Some species have lungs but retain gills throughout life (sirens and mudpuppies), others with lungs retain only the gill slits (Hellbender and Amphiuma), and still others (family Plethodontidae) lack both lungs and gills and respire mainly via the skin. Most salamanders feed rather indiscriminately on a wide range of invertebrates and other small prey, but some show strong preferences—for example, the Spring Salamander prefers smaller salamanders.

The order Caudata includes over 550 currently recognized species. Many families inhabit the north temperate zone, but only one family (Plethodontidae) is successful in the tropics (New World). The Carolinas and Virginia have a rich salamander fauna containing at least 78 currently recognized species. Most occur in the Mountains, unlike the other amphibians and reptiles, which have their greatest species diversity in the Coastal Plain. Our salamanders generally have small geographic ranges and a high degree of endemism. At least 11 species are probably endemic, and several others have over two-thirds of their ranges within the area. Only one species—the Eastern Newt—occurs virtually areawide.

Most (62) of our salamanders are plethodontids—a dominant family with eight genera, three of which are large: *Plethodon* (29 species), *Desmognathus* (17), and *Eurycea* (10). The only other large group is the Ambystomatidae, a family with seven species in our area, all in the genus *Ambystoma*. Nonetheless, our fauna is taxonomically diverse; three families and also five genera are each represented by only a single species.

Northern Dwarf Siren

Pseudobranchus striatus

100 to 175 mm (4 to 7 in.) These small, slender, eellike salamanders have external gills and tiny forelegs but no hind legs. Unlike other sirens, the dorsum and sides of the body are brown and conspicuously striped with yellow. The venter is grayish green with many yellow flecks. Dwarf Sirens also have a smaller and slimmer body than other sirens, a more pointed snout, 3 digits per limb, and only one pair of gill slits.

These unusual amphibians inhabit the southern half of the Coastal Plain in South Carolina. They find food and shelter in mud or amid thick aquatic vegetation in swamps, marshes, ditches, or ponds. They may be found in all months of the year.

Dwarf Sirens feed on small, slow-moving, or dead organisms, including aquatic insect larvae, crustaceans, small snails, and worms. Females lay fewer than 20 eggs in late winter or spring, attaching them singly to vegetation. During drought, Dwarf Sirens may become encased in dried mud and survive for many months. Under such conditions the gills shrink, but they can attain normal size within a week after the return of water.

Much remains to be learned about the biology of this infrequently encountered species in our area. It is listed as a species In Need of Management in South Carolina.

Lesser Siren *Siren intermedia*
150 to 380 mm (6 to 15 in.) This large, eellike salamander lacks hind limbs but has small front limbs (each with 4 digits) and bushy external gills. Costal grooves on each side between the front legs and cloaca number 31 to 35. The dorsum is black or brown, sometimes with minute black dots; the venter is slightly paler. This species is smaller and more slender than the Greater Siren and has fewer costal grooves.

Lesser Sirens inhabit most of the Coastal Plain in our region, but populations in northeastern North Carolina and eastern Virginia are scattered. They prefer the quiet, weed-choked waters of swamps, ditches, and ponds. They can be active in all months of the year, but peak feeding season is June and July.

Sirens communicate with one another by a series of clicking sounds made by rapid snapping of their keratinized jaws. They are predators on aquatic insects, snails, small clams, amphipods, crayfish, earthworms, and sometimes their own eggs. In late winter to early spring, females lay 100 to 1,500 eggs in small depressions on the bottom of a pond. Hatchlings are 11 mm (0.4 in.) long when they appear in late April and May. The body and head soon become boldly marked with a mid-dorsal, a lateral, and a ventrolateral light stripe. These stripes turn orange or red, only to disappear within a year. The stripe on the side of the snout is the last to disappear.

Greater Siren *Siren lacertina*
510 to 950 mm (20 to 37.5 in.) This
very large, eellike salamander lacks
hind limbs but has small front
limbs (each with 4 digits) and large
external gills. The dorsum is olive
to light gray, sometimes with black
spots, and often with small yellow or
white spots in juveniles. The venter
is bluish gray, sometimes with many
small, greenish yellow flecks or
spots. Larger size and more costal
grooves between the cloaca and
front limbs (37 to 40) distinguish
this species from the Lesser Siren.

Greater Sirens inhabit the

Coastal Plain of Virginia and the
Carolinas. They are nocturnal and
favor muddy and weed-choked
ditches, swamps, and ponds, as well
as lakes, rivers, and streams, where
they hide in burrows in the substrate. They are nocturnal and may
be active year-round except in the
coldest weather.

Greater Sirens eat crayfish, mollusks, worms, arthropods, and large
amounts of algae (which may be
ingested accidentally). Extensive fat
deposits enable sirens to survive several years without feeding, as when
stranded by drought in underground
chambers or burrows. Each female
lays about 500 eggs in small groups
scattered about on the bottom.
Hatchlings measure 16 mm (0.7 in.)
in total length and during the first
year of life have a red or yellow stripe
on each side of the body and another
on the dorsal fin.

Sirens possess so many features
not found in other amphibians that
many systematists advocate placing
them in a separate order.

Hellbender

Cryptobranchus alleganiensis

300 to 740 mm (12 to 29 in.) "Impressive" best describes this large, fully aquatic salamander. Adults and juveniles are gray, brown, or yellowish brown, often with irregular dark spots or blotches. The wide, flat head bears small, lidless eyes. The legs are short and stout, and a large wrinkled fold of skin extends along each side of the body. Adults lack external gills but have a small gill slit on each side of the throat.

Hellbenders live mainly in streams of the Mississippi drainage in southwestern Virginia and western North Carolina. There are no recent records from South Carolina; two records from Tugaloo Lake in Oconee County are thought to represent introductions. These salamanders require clean, cold water and prefer large, clear, fast-flowing streams with big, flat rocks, under which they often hide.

In late summer a female lays a string of 300 to 400 eggs in a depression dug by a male under a rock. Several females may deposit in the same nest, and choice nesting rocks may be used repeatedly over many years. The male guards the eggs aggressively until they hatch into larvae about 30 mm (1.2 in.) long in late summer to early fall. Larvae lose their gills when they are about 100 to 130 mm (4 to 5 in.) long and 18 months old. Hellbenders eat mostly crayfish, earthworms, and aquatic insects. Active day and night, they are sometimes captured by anglers using live bait. Contrary to folklore, Hellbenders are harmless to humans. Long-lived animals, they are excellent indicators of stream quality.

The Hellbender has undergone serious declines over much of its range in recent years. Threats include siltation, dams, chemical pollutants, overcollecting, and senseless killing of individuals. It is listed as a species of Special Concern in North Carolina.

Neuse River Waterdog
Necturus lewisi

170 to 276 mm (6.5 to 11 in.) Waterdogs differ from other large salamanders by having 4 toes on each hind foot and conspicuous external gills that are retained throughout life. This species differs from the Common Mudpuppy in having a lighter dorsal color with more pattern contrast and a more heavily spotted belly. Juveniles lack the dark stripe on the back characteristic of that species. Dark spots form gradually in juveniles, first on the back and then, as maturity approaches,

on the belly. The head is more depressed and the body is stockier than in the Dwarf Waterdog.

This species is endemic to North Carolina. It inhabits the main streams and larger tributaries of the Neuse and Tar rivers from well above tidewater into the lower Piedmont. It prefers leaf beds in quiet water in winter but is only infrequently found in this habitat in summer. Prey include a wide variety of insects and other arthropods, mollusks, earthworms, leeches, and small vertebrates.

Females bear large-yolked eggs in April and May and attach them beneath large, flat rocks in water over 0.3 m (1 ft.) deep. Clutch sizes ranged from 19 to 36 eggs in 8 documented nests. Hatching occurs in July. Sexual maturity is attained at an age of about 6 years for females and 5 years for males. Some are thought to live as long as 18 years.

The Neuse River Waterdog is listed as a species of Special Concern in North Carolina.

Common Mudpuppy
Necturus maculosus

203 to 228 mm (8 to 9 in.) Adults of this permanently aquatic salamander are brown to gray with scattered black spots or blotches on the body and tail. The belly is cream to white with scattered black markings. There are 4 toes on each rear foot, and the external gills are large, bushy, and red. Juveniles have a dark stripe along the middle of the back that is flanked on each side by a light yellow stripe and a dark band below.

In our area, this species is known only from the Holston and Clinch river systems in Virginia and from the French Broad and New river systems in North Carolina; but its distribution is not well known, and it may occur in some other drainages. Common Mudpuppies are most readily found in late fall or early winter when they aggregate in leaf beds in slack water. They eat crayfish, aquatic insects, worms, snails, small vertebrates, and their own eggs.

Breeding has not been observed in our area, but farther north, about 30 to 140 eggs are laid in May and June on the undersides of large rocks in shallow, slow-moving water in lakes and streams. Females guard their eggs from predators such as crayfish and fish. Hatching occurs in midsummer.

The Common Mudpuppy is listed as a species of Special Concern in North Carolina.

Dwarf Waterdog
Necturus punctatus

115 to 190 mm (4.5 to 7.5 in.) Most Dwarf Waterdogs are uniformly dark brown above and pale below; however, in the Cape Fear and Lumber systems, most adults are distinctly spotted, especially on the tail. The spots are about the same size as the eye—smaller than in the Neuse River Waterdog. The Dwarf Waterdog is also smaller, more slender, and more cylindrical.

This species occurs in the Coastal

Plain and eastern Piedmont from the Chowan and Meherrin river drainages in south central Virginia to the Ocmulgee-Altamaha system in Georgia. It inhabits smaller streams than does *N. lewisi* and is found down to the upper limits of tidewater. Along the Fall Line in the Tar and Neuse rivers of North Carolina the two species occur together. Juveniles burrow in the silty bottoms of streams. Adults congregate in leaf beds in winter and are apparently relatively inactive in summer.

Adults and juveniles include worms, mollusks, crustaceans, aquatic insects, and other salamanders in their diet. Maturity is reached a year earlier than in *N. lewisi*. Large specimens are probably at least 10 years old. Females lay 20 to 40 eggs in spring and presumably attach them to the undersides of submerged logs and other debris, like other waterdogs. Although this species is common in some streams, its life history is poorly known.

Two-toed Amphiuma
Amphiuma means

460 to 1,162 mm (18 to 46 in.) This robust, eellike salamander—North America's longest amphibian—has 2 pairs of tiny legs with 2 toes on each foot. The body is uniformly dark brown, dark gray, or black above and lighter gray below. The superficially similar sirens have bushy, external gills and a single pair of anterior legs.

This relatively common but infrequently encountered species inhabits streams, abandoned rice fields, ditches, and shallow ponds in pine savannas, hardwood forests, and swamps. It occurs throughout the Coastal Plain and in portions of the adjacent Piedmont from east of Richmond, Virginia, south through the Carolinas. Amphiumas may travel between wetlands over land during heavy rains.

Two-toed Amphiumas are most active at night. Their diet includes insects, worms, crayfish, snails, small vertebrates, and carrion. They may bite defensively, and the teeth of large individuals can inflict serious lacerations; handle them with caution! The life history of this species is poorly known. In winter months, females deposit long, rosarylike strings of 10 to 350 eggs in depressions beneath logs, boards, or other objects in moist or wet areas, including alligator nests. Females remain with the eggs, which hatch about 5 months later into aquatic larvae averaging about 55 mm (2.2 in.) long. Recently transformed young are about 70 mm (2.75 in.) long.

Flatwoods Salamander
Ambystoma cingulatum
90 to 129 mm (3.5 to 5 in.) This dark salamander has grayish reticulations on the dorsum and sides and light flecks on the venter. It differs from other *Ambystoma* in our area by its small size, slender body with a relatively small head, distinct reticulated markings, and 15 costal grooves (rather than 11 to 13). Mabee's Salamander is also small but is dark brown with light lateral flecks.

In our region, Flatwoods Salamanders occur in the southern half of the Coastal Plain of South Caro-

lina. The chief habitat was flatwoods dominated by Longleaf Pine and Wiregrass, but deforestation has changed much of it to Slash Pine forests. These increasingly rare salamanders may be found beneath logs near cypress ponds, swamps, and pitcher plant bogs.

Most breeding occurs in November in our area. About 100 to 220 eggs are deposited in small groups on the ground under logs and other debris near Pond Cypress or cypress-tupelo ponds. The eggs are unattended and hatch a few weeks later when the site is flooded by rain. The aquatic larvae are very colorful: black with conspicuous light longitudinal stripes that provide concealment in aquatic grasses. Metamorphosis occurs in March or April at a total length of about 70 mm (2.75 in.). Larvae and adults prey on a wide variety of invertebrates.

The Flatwoods Salamander is federally listed as Threatened and state-listed as Endangered in South Carolina.

Jefferson Salamander
Ambystoma jeffersonianum

110 to 220 mm (4.25 to 8.5 in.) This dark brown or gray salamander is more slender than most species of *Ambystoma* and has long, slim toes and a laterally compressed tail. It is marked on the sides with bluish flecks, which are most apparent on juveniles. Members of the Slimy Salamander complex may look superficially similar but have round tails and nasolabial grooves.

Jefferson Salamanders inhabit mixed hardwoods from New England south, mainly in the Valley and Ridge physiographic province of Virginia to the Clinch River. They occur in valleys and on mountains to about 900 m (3,000 ft.) elevation.

With the first warm rains of early spring they migrate to woodland ponds, borrow pits, and sinkholes to court and spawn. Males deposit gelatinous spermatophores on debris in water, and with her vent the courted female nips off the sperm-bearing cap to achieve internal fertilization. She then deposits several egg masses, each containing 2 to 70 eggs, along twigs and stems. The larvae hatch in about a month at about 12 mm (0.5 in.) and transform in July or August at about 50 to 75 mm (2 to 3 in.) total length. Larvae consume virtually any aquatic prey they can catch and swallow—including one another. Adults feed on worms, snails, insects, and other prey while in underground burrows outside the breeding season.

Mabee's Salamander
Ambystoma mabeei

80 to 114 mm (3 to 4.5 in.) This brownish gray salamander has a relatively small head and long, slender toes. Silvery white flecks are abundant along the sides but sparse on the back. Slimy Salamanders may look similar but are black with naso-labial grooves and round tails. Mole Salamanders can also be brownish with light flecking but are much stockier and have larger, broader heads. Flatwoods Salamanders are dark gray, more slender, and have more light flecks on the body.

The range of this species in-cludes most of the Coastal Plain in the Carolinas northward to New-port News, Virginia. Mabee's Sala-mander is characteristically a pine savanna and mixed pine-hardwood species, living in burrows near the edges of fish-free ephemeral ponds, sinkholes, Carolina bays, and other depressional wetlands. It also occurs in low, wet woods and swamps.

The breeding season extends from late fall to early spring. Follow-ing mating in ponds during warm rains, females attach their eggs singly or in loose chains of 2 to 6 to leaves, twigs, or other bottom debris in shallow water. The aquatic larvae hatch after 9 to 14 days and average 8.5 mm (0.3 in.) long. Transforma-tion occurs in April to early June at sizes of 50 to 60 mm (2 to 2.3 in.). Larvae eat mostly small aquatic invertebrates; various arthropods and snails have been documented as prey of adults.

Mabee's Salamander is listed as a Threatened species in Virginia.

Spotted Salamander
Ambystoma maculatum

150 to 249 mm (6 to 10 in.) Bright yellow, round spots in two irregular rows on a dark background identify this stout-bodied species. Tiger Salamanders may look similar, but their light markings are more oval and less regular in position. Individuals from the Mountains average 25 mm (1 in.) longer than those from the Piedmont.

Spotted Salamanders inhabit deciduous forests and breed in semipermanent pools ranging from road rut puddles to large sinkhole ponds. They usually avoid bottomlands subject to regular flooding and permanent ponds containing fish. They are common in the Piedmont and Mountains. Populations are scattered but locally common in the Coastal Plain. Because of their fossorial habits they are rarely found except during the short breeding season.

With the first warm rains in winter to early spring, these salaman-

ders leave their burrows and migrate by night to the breeding ponds.

Up to 350 eggs are laid in clear to opaque, jellylike masses attached to sticks and stems in the water. Adults return to their burrows with the next rain. The eggs hatch in 30 to 55 days. Larvae transform between June and September, depending on location. They acquire the adult pattern a few days or weeks later and begin their fossorial life, reaching maturity in 2 to 3 years for males and 3 to 5 years for females. Larvae and adults feed mostly on insects and other invertebrates.

Marbled Salamander
Ambystoma opacum

90 to 127 mm (3.5 to 5 in.) This stocky salamander is shiny black with a series of conspicuous white or light gray crossbands over the back. These are brighter and more distinct in males, especially during the breeding season, and may be partially or completely fused to form 2 stripes along the back. No other salamander in our area has such markings.

Marbled Salamanders occur throughout most of the Carolinas and Virginia but are uncommon and local in the Mountains. They reside under logs or rocks, usually in floodplains or in moist, sandy areas along ponds and streams.

Unlike most species in the genus, Marbled Salamanders breed in early fall, on land adjacent to ephemeral ponds and depressions. Each female lays about 100 eggs in a small depression in the soil beneath leaves or a log; she usually remains with the eggs until they hatch after winter rains inundate the nest sites to form temporary ponds. The larvae are carnivorous, eating small aquatic invertebrates (and sometimes one another). They transform in April or May, depending on elevation and latitude. Adults are more readily found throughout the year than are other *Ambystoma*. They eat earthworms, insects, and snails.

Mole Salamander
Ambystoma talpoideum

80 to 122 mm (3 to 5 in.) This gray, brown, or dark brown salamander has a large head; a short, stocky body; and relatively large legs. Scattered bluish white flecks occur on the back and sides. The belly is bluish gray with light flecks, except in recently transformed young, which have a dark stripe down the middle, a remnant of the larval pattern.

Mole Salamanders occupy underground burrows in pine savannas, hardwood forests, and margins of swamps. They are rarely encountered except during the breeding season, when they congregate in shallow ponds. Their unusual distribution includes the southern half of the South Carolina Coastal Plain and several disjunct localities in the Piedmont and southern Mountains of North Carolina and the western Piedmont of Virginia.

Breeding typically occurs in midwinter in Coastal Plain locations and in late winter in Piedmont populations. Females deposit 200 to 500 eggs singly or in small, loose clusters of 10 to 40 eggs each, attached to stems of vegetation or other objects in shallow ponds. The larvae usually transform in the summer and fall but may overwinter; individuals in some populations attain sexual maturity in the larval form before transforming. The newly transformed young are 55 to 70 mm (2.25 to 2.75 in.) long. Prey of adults and juveniles include a wide array of invertebrates. Adults are known to live as long as 8 years.

The Mole Salamander is listed as a species of Special Concern in North Carolina.

Eastern Tiger Salamander
Ambystoma tigrinum

178 to 279 mm (7 to 11 in.) This large salamander has yellow, yellowish brown, or olive spots against a background of dark brown or black. The irregular spots extend well onto the sides. The belly is yellowish and marbled with darker pigment. Spotted Salamanders are similar but have two distinct rows of yellow spots on the back and yellow or reddish orange spots on the head.

This rare species in our area is known only from widely scattered localities in the Coastal Plain, Sandhills, and Piedmont and from one

locality in the Shenandoah Valley in Virginia. It inhabits burrows in sandy areas and breeds in upland ephemeral wetlands such as shallow ponds, Carolina bays, borrow pits, and sinkholes, chiefly in Longleaf Pine savannas and hardwood forests on alluvial slopes.

Males emerge from underground retreats and enter ponds before females in fall or winter, sometimes when snow and ice are present. Freshly deposited eggs have been found from early November to mid-March. The loose, globular, or oblong egg cluster contains from 15 to 200 (usually about 50) eggs and is attached to stems in shallow ponds. The larvae transform from May to July at a total length of about 70 to 120 mm (2.75 to 4.75 in.). Larvae eat aquatic invertebrates and amphibian larvae. Terrestrial adults eat insects, worms, and small vertebrates. They have been known to live for more than 20 years.

The Eastern Tiger Salamander is listed as Endangered in Virginia and Threatened in North Carolina.

Eastern Newt (adult)

Eastern Newt
Notophthalmus viridescens
60 to 140 mm (2.5 to 5.5 in.) The aquatic adults of this interesting and variable salamander are yellowish brown, olive green, or dark green above, and yellow with small black spots below. Red spots encircled with black occur on the dorsum in most populations, but newts in Coastal Plain populations from the southern half of North Carolina to the Santee River in South Carolina have broken red stripes with black borders. Those from south of the Santee lack or have reduced red markings. Terrestrial subadults (efts) have thicker, rougher skins and vary from reddish brown to bright orange-red. Newts lack costal grooves.

Adults live in ponds, lakes, and pools near rivers and streams, and efts inhabit moist, forested areas. This species is common to abundant throughout most of our area.

An elaborate courtship occurs in spring and fall. Females deposit eggs singly on leaves of submerged plants in ponds or lakes in late winter, spring, and early summer. They often fold a leaf around each egg, effectively hiding it from view. Hatching occurs after a developmental period of up to 35 days, and the newly emerged larvae average 7.5 mm (0.3 in.) long. Transformation of the larvae into terrestrial efts takes place in summer or fall, but in some

Eastern Newt (eft)

areas of the Coastal Plain, under certain conditions, the eft stage is omitted. Newly transformed young are about 36 to 41 mm (1.4 to 1.6 in.) long. The bright coloration of efts warns visually oriented predators of their skin toxins. They are often seen abroad on the surface, especially on wet or overcast days. They return to ponds when they reach maturity in 1 to 8 years. The details of the life cycle vary considerably throughout the area and over the entire range. Newts feed on a variety of aquatic invertebrates, including insects, crustaceans, and mollusks. They also eat the eggs of other amphibians.

Green Salamander *Aneides aeneus*
80 to 140 mm (3 to 5.5 in.) This dorsoventrally flattened salamander has green or yellowish green, lichen-like patches on a background of dark brown or gray. The belly is pale yellowish white. The limbs and tail are relatively long, and the toes on the large feet are webbed and have expanded tips.

Green Salamanders inhabit moist crevices on shaded quartzite and limestone rock outcrops and cliff faces in hardwood forests. Recent research has revealed that they may be seasonally arboreal, frequently making extensive use of tree trunks and sometimes ascending into the canopy. In our area, this now scarce species is known only from scattered localities in the Mountains of southwestern Virginia, southwestern North Carolina, and northwestern South Carolina.

In May to July, females deposit clusters of 10 to 26 eggs on the roofs of horizontal crevices and the sides of vertical crevices. They remain with the eggs throughout the development period of 84 to 91 days. Hatching occurs in September and October. The newly hatched young average about 20 mm (0.8 in.) long and are miniature replicas of the adults. These salamanders hibernate in deep crevices from November through late March. On warm, rainy nights they frequently forage on rock faces for insects, spiders, and other arthropods.

The Green Salamander is listed as Endangered in North Carolina.

Seepage Salamander
Desmognathus aeneus

44 to 57 mm (1.75 to 2.25 in.) This tiny salamander has a reddish bronze dorsal stripe with a series of dark, irregular spots or a dark line down the middle of the back. Chevronlike marks are sometimes present. There is usually a Y-shaped mark on the head and a dorsal, yellowish or reddish spot on the thigh. The belly is heavily mottled with dark pigment. Pigmy Salamanders and young Ocoee Salamanders are similar but generally have plain bellies. Pigmy Salamanders are also stockier and have a dorsal, yellowish

or reddish spot on the upper arm and the thigh and silvery pigment on the lower sides of the body.

Seepage Salamanders spend most of their lives beneath leaf litter near seepages, springs, or streams in hardwood forests. They seldom appear on the surface, possibly because their small size makes them vulnerable to many predators. In our area, they occur in the southwestern corner of North Carolina and in Oconee County, South Carolina.

In late April to early May, females deposit compact clusters of 11 to 14 eggs beneath moss or decomposing logs in seepages or springs. They remain with the eggs, which hatch in late May to early August. Hatchlings are essentially "terrestrial larvae," resembling the adults. They may emerge with small gills, but these are lost within a few days or weeks; there is no true aquatic larval stage. Seepage Salamanders eat mostly tiny invertebrates, including mites, spiders, isopods, centipedes, millipedes, worms, snails, pseudoscorpions, beetles, flies, and springtails.

Southern Dusky Salamander
Desmognathus auriculatus

80 to 163 mm (3 to 6.5 in.) This dark brown to almost black salamander has two rows of conspicuous white or reddish orange spots on each side of the body. The laterally compressed tail is trigonal in cross section at the base, is keeled, and has a middorsal reddish orange stripe. The belly is dark brown and heavily peppered with white or pale yellow. Some Northern Dusky Salamanders are similar but have lighter bellies, less conspicuous lateral light spots, and more tapered tails. Larvae of the two species are more easily distinguished. The gills of Southern Dusky Salamander larvae are bushy and pigmented and have 30 to 43 filaments on each side; those of the Northern Dusky are less bushy and glistening white and have 19 to 33 filaments on each side.

Southern Dusky Salamanders live under leaf litter and decomposing logs in swamps and bottomland forests throughout much of the Coastal Plain in southeastern Virginia and the Carolinas.

Females deposit clusters of 9 to 26 eggs in cavities beneath moss and bark or within logs near water in July to October. Like other females in this group, they remain with the eggs until they hatch into aquatic larvae in the fall. Transformation to the juvenile stage occurs in late spring. Their diet consists of a wide array of insects, spiders, worms, and other small invertebrates.

Relationships within the genus *Desmognathus* are not fully understood. Recent research indicates that populations traditionally regarded as *D. auriculatus* over most of our area are more genetically similar to *D. fuscus* and *D. carolinensis*. *Desmognathus auriculatus* may not, in fact, occur within our region at all. Animals currently treated as *D. auriculatus* in Virginia and the Carolinas may instead represent several cryptic, undescribed species.

Carolina Mountain Dusky Salamander

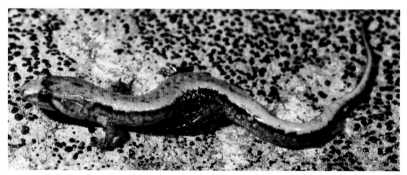

Allegheny Mountain Dusky Salamander

MOUNTAIN DUSKY
SALAMANDER COMPLEX

Carolina Mountain Dusky
Salamander
Desmognathus carolinensis

Allegheny Mountain Dusky
Salamander
Desmognathus ochrophaeus

Ocoee Salamander
Desmognathus ocoee

Blue Ridge Dusky Salamander
Desmognathus orestes
70 to 111 mm (3 to 4.5 in.) The salamanders in this confusing species
complex were until recently regarded
as a single species (*D. ochrophaeus*),
but they differ genetically. All are

relatively slender, medium-sized
Desmognathus with rounded, un-
keeled tails slightly longer than their
bodies. All species in this complex
show extensive individual variation.
Most young animals are spotted
dorsally; but many adults are drab
and nondescript, while others have
bright colors and patterns. Aside
from molecular differences, they are
best distinguished by geographic
range.

In our region, *D. ochrophaeus*
primarily occupies the higher ridges
of the Valley and Ridge province
in western Virginia. This species
tends to have straighter dorsolateral
stripes than other members of the
complex.

Desmognathus orestes occurs in the southern Blue Ridge province of Virginia and the northwestern corner of North Carolina.

Desmognathus carolinensis occurs in the central North Carolina Mountains between about Linville Falls and McKinney Gap in Burke and Mitchell counties and the Pigeon River valley in Buncombe and Haywood counties. Some individuals, particularly those from Mount Mitchell, are larger and darker than other members of the complex, with short tails and small heads.

Desmognathus ocoee occupies the remainder of the Mountains in southwestern North Carolina and northwestern South Carolina and is probably the most variable member of this complex. Some specimens in the Nantahala and adjacent Blue Ridge Mountains may have bright orange or red patches on their cheeks or upper legs; these are possible Batesian mimics of the less palatable Jordan's and Sherman's salamanders.

Members of this species complex differ from most other *Desmognathus* in being more terrestrial, often more brightly colored, and more variably marked. The tail is also longer, rounder, and more tapered and usually lacks a dorsal keel. Males average slightly larger than females. They are uncommon at low elevations, where they inhabit seepages and stream margins. At high elevations they range away from streams, often living under rocks, leaves, and logs in adjacent woodlands. Among various populations, breeding may take place from late winter through late summer. Females lay 8 to 37

Blue Ridge Dusky Salamander

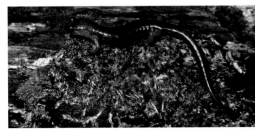

Ocoee Salamander

eggs, often attaching them to the undersurfaces of rocks or logs along seeps and streams, and attend them until they hatch. The aquatic larvae transform in about 2 to 10 months, depending on when the eggs were laid. Insects, earthworms, and other small invertebrates are taken as prey.

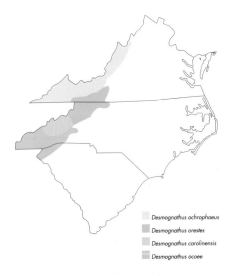

Desmognathus ochrophaeus
Desmognathus orestes
Desmognathus carolinensis
Desmognathus ocoee

Spotted Dusky Salamander
Desmognathus conanti

64 to 127 mm (2.5 to 5 in.) This salamander resembles the Northern Dusky Salamander but has 5 to 8 pairs of red to golden spots or blotches on the dorsum of the brown to gray body. The venter is cream colored and usually patternless. The base of the tail is somewhat compressed side to side and has a prominent keel. The Spotted Dusky and Northern Dusky overlap in portions of the Carolinas, thus making

identification difficult. Adults of the latter species lack strong spotting on the back. *Desmognathus conanti* is sometimes regarded as a subspecies of *D. fuscus*.

Spotted Dusky Salamanders occur in far southwestern North Carolina and much of western South Carolina. Their overall range extends from Kentucky to Louisiana. These salamanders occupy seepage habitats and streams in deep, forested ravines and margins of slow-moving, muddy streams and swamps.

From July to October, females lay 13 to 24 eggs under rocks, leaf mats, moss, decomposing logs, and debris on the edge of seepages and streams. The larvae hatch in about 5 to 7 weeks and may remain in the larval stage for another 12 to 13 months. Prey include a variety of arthropods, worms, and other invertebrates. The life history and ecology of this species are poorly known and have not been studied in our region.

Dwarf Black-bellied Salamander
Desmognathus folkertsi

115 to 150 mm (4.5 to 5.9 in.) This recently described salamander was long overlooked due to its close resemblance to the larger and sympatric Black-bellied Salamander, from which it differs genetically. Dwarf Black-bellied Salamanders have vermiculate dorsal patterns of brown and black. On some, the brown markings are arranged as blotches similar to those of Shovel-nosed Salamanders. Black-bellied Salamanders are usually more uniformly brown or dark greenish with a few irregular black spots, and they tend to have slightly longer limbs and toes and taller, narrower tails. In adults of both species, the belly is black. The larvae are similar in appearance.

The Dwarf Black-bellied Salamander is known in our area only from southeastern Clay County, North Carolina, near the Georgia state line. It is otherwise known only from a few counties in northern Georgia, but it is suspected to be more widespread.

Much remains to be learned about the ecology, life history, and distribution of this salamander. Its habits are presumably similar to those of the Black-bellied Salamander. Natural nests are unknown, but clutch sizes estimated from developing follicles in mature females ranged from 2 to 62 (average 35 to 40). The larval period is thought to last 2 years. Aquatic insects have been reported as prey.

Northern Dusky Salamander
Desmognathus fuscus
60 to 141 mm (2.5 to 5.5 in.) The young of this highly variable sala-mander usually have 5 to 8 pairs of buff-colored, yellow, or reddish orange dorsal spots on a background of dark brown or gray. Most older individuals have a wide dorsal stripe with dark, wavy or scalloped edges extending to the base of the tail. The stripe on the top of the tail is usually reddish. Some specimens may be uniformly dark gray or brown. The belly is usually mottled lightly with gray or brown. The tail is keeled and compressed side to side. Salaman-ders in the Mountain Dusky com-

plex have round tails that taper to a filament. See also the account of the Southern Dusky Salamander.

Northern Dusky Salamanders are abundant in streams, springs, and seepages in bottomland forests and wooded ravines in the Piedmont, portions of the upper Coastal Plain, and many areas in the Mountains of the Carolinas and Virginia.

Females deposit compact clus-ters of about 11 to 36 eggs in cavi-ties of decomposing logs, in stream banks or seepages, or on the under-surfaces of rocks in streams. They re-main with the eggs until they hatch into aquatic larvae in July to Septem-ber. Transformation to the juvenile stage occurs late in the following June or July. These salamanders eat a variety of arthropods, worms, and snails. Adults will cannibalize juve-niles, and females may eat the eggs of other females as well as their own.

Salamanders in the genus *Des-mognathus* are a confusing group, and their systematic relationships are not well understood. *Desmogna-thus fuscus* may represent a complex of several closely related species.

Imitator Salamander
Desmognathus imitator

70 to 110 mm (3 to 4.5 in) This cryptic, highly variable species resembles the Ocoee Salamander in size and general appearance but differs in genetic makeup. Many populations and individuals can be reliably distinguished from *D. ocoee* only by molecular analysis. Only certain populations along the main ridge of the Great Smoky Mountains can be distinguished by their color patterns. Most *D. imitator* from these populations have a wide, indistinct dorsal stripe with strongly undulating edges. In contrast, most *D. ocoee* have a distinct wide dorsal stripe of light yellow to dark brown bordered along each side by a straight to moderately undulating black stripe. Many individuals of both species darken with age and become almost black and patternless. Many *D. imitator* have red, orange, or yellow cheeks and are perhaps Batesian mimics of the more unpalatable Jordan's Salamander. Some *D. ocoee* also have colored cheek patches, but those from within the range of *D. imitator* do not. To distinguish *D. imitator* from *Plethodon jordani*, look for these desmognathine features: (1) a pale diagonal line, or trace of one, from the eye to the angle of the jaw, (2) a bent-head profile, and (3) hind legs much larger than forelegs.

Imitator Salamanders are endemic to the Great Smoky Mountains National Park area of North Carolina and adjacent Tennessee. They occur under rocks, logs, and leaf litter in cool, moist spruce-

fir and hardwood forests. Mostly nocturnal, they feed on a variety of small invertebrates. They prefer to remain near intermittent streams and seeps, whereas Ocoee Salamanders are often found on the forest floor, far from water. Female Imitator Salamanders attach their eggs to the undersides of rocks in seepages in spring and early summer. Other aspects of this salamander's life history and ecology are poorly known. Populations inhabiting rock faces in the vicinity of Waterrock Knob in Haywood and Jackson counties, North Carolina, differ in appearance from other populations and may represent an undescribed form.

Shovel-nosed Salamander
Desmognathus marmoratus

88 to 143 mm (3.5 to 5.5 in.) The dorsum of this highly aquatic salamander is usually dark brown with two rows of light blotches. Typically, a light area extends from the eye to the angle of the jaw, as occurs in other *Desmognathus*. The venter is usually dark gray with a whitish central area. Individuals from the Nantahala River drainage are especially dark. An interesting type of depigmentation

occurs especially in southern populations. These individuals are mostly whitish yellow but may have dark blotches. Shovel-nosed Salamanders are often confused with Black-bellied Salamanders but are more aquatic and have broader tail fins, flatter snouts, and slitlike internal nostrils. They were formerly assigned to a separate genus (*Leurognathus*).

Shovel-nosed Salamanders inhabit trout streams and small brooks in the Mountains from southwestern Virginia to northeastern Georgia. They spend much time under rocks and feed mainly on aquatic insects.

Females lay 20 to 65 eggs in May and June. Eggs are attached to the undersides of rocks in the main current. The female stays with her eggs throughout the 2.5 months of development. The larvae are about 20 mm (0.75 in.) in length when they hatch. They metamorphose 10 to 20 months later when 40 to 65 mm (1.5 to 2.5 in.) long.

Seal Salamander

Desmognathus monticola
80 to 149 mm (3 to 6 in.) This robust, buff-colored, greenish gray, or light brown salamander has dark, wavy or wormlike marks on its back and a pale or very lightly mottled belly. The posterior half of the tail is keeled and laterally compressed. The young are greenish brown with 4 pairs of reddish orange spots. In the Virginia Blue Ridge, the dark dorsal markings are usually reduced to scattered, round dots. Northern Dusky Salamanders are smaller, less robust, and have more heavily mottled bellies and shorter tails.

Seal Salamanders are abundant along streams, seepages, and springs in the Mountains and portions of the adjacent Piedmont. They are active year-round except during the coldest weather.

In May and June, females deposit clusters of 16 to 40 eggs in cavities in decomposing logs along the edges of stream banks, under moss, or on the undersurfaces of rocks in streams. They remain with the eggs until they hatch into aquatic larvae in September. The larval period lasts 8 to 10 months, whereupon they lose their gills and transform into juveniles about 23 to 30 mm (1 to 1.2 in.) long. These salamanders eat aquatic and terrestrial insects, earthworms, and other invertebrates. Adults will forage on land during warm rains.

Virginia Dusky Salamander
Desmognathus planiceps
64 to 115 mm (2.5 to 4.5 in.) This species is closely related and similar to the Northern Dusky Salamander. It was originally described in 1955, subsequently regarded as a variant of the Northern Dusky Salamander for several decades, and recently recognized and "resurrected" as a distinct species. The body and tail are brown to gray with a variable pattern on the back usually consisting of 6 to 7 pairs of irregular spots and a yellowish to reddish brown dorsal stripe that may have wavy to straight edges. The belly is cream to light brown with some dark mottling. Patterns darken with age. The most reliable method of distinguishing this salamander from the Northern

Dusky Salamander is genetic analysis. Male Virginia Dusky Salamanders have noticeably broader teeth than the Northern Dusky Salamanders, but females of the former have narrower teeth than males; thus this characteristic only holds true with males. Rely in part on geographic distribution to help identify this species.

Virginia Dusky Salamanders are endemic to a small area in the southwestern Piedmont of Virginia, from the slopes of the southern Blue Ridge escarpment eastward through Patrick, Franklin, and Henry counties. They occupy mostly cool springs, seepages, and streams. They are not active during winter months but can be found in streams and other wetlands from about March to October. During dry summer months these salamanders will move to underground retreats.

In spring, early summer, or fall, females lay 11 to 36 eggs in crevices underground, in stream beds, or under rocks. Larvae either metamorphose in the fall or the following spring, depending on when the eggs were laid. Almost nothing is known of the behavior, life history, diet, or ecology of this species.

Black-bellied Salamander
Desmognathus quadramaculatus
100 to 210 mm (4 to 8 in.) This large, robust salamander has a brown or dark greenish gray back, a black belly, and a keeled tail that is shorter than the body. Two rows of conspicuous white spots are usually visible on each side of the trunk. In young subadults, the belly is unpigmented; older subadults have patches of black interspersed with areas of yellowish white. Although a dark Shovel-nosed Salamander is similar, it lacks the sharp contrast in color between the belly and the lower sides and has a flattened, wedge-shaped snout and smaller eyes.

Black-bellied Salamanders inhabit mountain streams and often rest on wet rock ledges near waterfalls, seepages, or springs. They occur in the Mountains throughout the Carolinas and southwestern Virginia and in portions of the adjacent Piedmont.

In May and June, females deposit clusters of 21 to 69 eggs on the undersurfaces of rocks or on tree roots in streambeds. They remain with the eggs until they hatch into aquatic larvae in late July to September. The larval period lasts 2 to 3 years. These predators eat a wide variety of invertebrates as well as smaller salamanders, including other members of their genus.

Santeetlah Dusky Salamander
Desmognathus santeetlah

64 to 95 mm (2.5 to 3.75 in.) This fairly recently described salamander is sometimes treated as a subspecies of *Desmognathus fuscus*, to which it is very similar, although slightly smaller and with a more subdued pattern. The two also differ genetically. Adults usually have a uniformly greenish or brownish dorsum with occasional reddish spots. Faint, interrupted dorsolateral stripes may be present. The belly, sides, and undersurface of the tail are usually washed with yellow, and there is a ventrolateral row of light "porthole" spots along each side of the body. The tail is weakly keeled. Older adults frequently have whitish flecking dorsally, which may be more pronounced on males. Juveniles usually have 5 pairs of chestnut spots on the back between the front and hind limbs; these may fuse to form a wavy stripe.

This salamander is endemic to the Great Smoky, Unicoi, and Balsam mountains of North Carolina and Tennessee, where it inhabits headwater streams and seeps at elevations from about 675 to 1,805 m (2,220 to 5,920 ft.).

Most nesting probably occurs from mid-May to early July. Females lay 10 to 35 eggs, most often beneath mosses growing on rocks or logs in seepage areas, and attend them during the 45- to 60-day incubation period. Santeetlah Dusky Salamanders eat small insects and other invertebrates. Much remains to be learned about their ecology and life history.

Black Mountain Salamander
Desmognathus welteri

80 to 170 mm (3 to 6.5 in.) This large, nondescript salamander is brown, usually with scattered dark spots, but occasional specimens are uniformly tan above. The venter is yellow and lightly mottled with dark pigment, and the toes are dark-tipped. Sympatric Northern Dusky Salamanders have dorsolateral dark stripes, a reddish middorsal stripe above the base of the tail, and light toe tips. Seal Salamanders have a dorsal pattern of dark brown to black spots or bars, remnants of the rims of the juvenile spots, and pale and evenly pigmented bellies. The Black Mountain Salamander is also more aquatic than the other *Desmognathus* within its range.

This species is known from the Cumberland Mountain and Cumberland Plateau region of southeastern Kentucky and southwestern Virginia. Adults prefer large, turbulent brooks, living under stones and in crevices in the splash zone. Larvae and juveniles inhabit spring seep tributaries.

About 18 to 33 eggs per clutch are laid in June. Females usually coil about their eggs during the incubation period. The larvae hatch in September, and the larval period lasts nearly 2 years, whereas that of the Northern Dusky Salamander and Seal Salamander is less than a year. Black Mountain Salamanders are known to eat earthworms and a variety of insects.

Pigmy Salamander
Desmognathus wrighti

38 to 51 mm (1.5 to 2 in.) This diminutive salamander usually has a
reddish bronze dorsal stripe with a
series of chevronlike marks along
the middle of the back, silvery pigment on the lower sides of the body,
an unpigmented belly, and a short
tail. The snout and eyelids are moderately to conspicuously rugose.
The similar Seepage Salamander
has a pigmented belly, a dark stripe
or series of dots on the back, and a

longer tail; its snout and eyelids are
usually smooth, and chevrons, if
present, do not extend the width of
the dorsal stripe.

Pigmy Salamanders live under
moss, leaf litter, rotten logs, bark on
stumps, or rocks in high-elevation
spruce-fir forests, but they move into
seepage areas in winter. Some populations occur in hardwood forests
at lower elevations. This species
is endemic to the high mountains
of southwestern Virginia, western
North Carolina, and adjacent Tennessee.

Pigmy Salamanders are active
chiefly on dark, humid nights and
may climb trees to heights of about
2 m (6.5 ft.) to forage on insects. In
late summer and fall, females deposit clusters of 8 to 10 eggs in an
underground cavity near a seepage
or stream. There is no larval period;
all development occurs in the egg.
Hatchlings appear in October and
have conspicuous spots but otherwise resemble adults.

Northern Two-lined Salamander
Eurycea bislineata

64 to 121 mm (2.5 to 4.75 in.) This slender salamander has a yellow to reddish orange dorsal stripe with black dots or flecks. A dark dorsolateral stripe extends to at least the middle of the tail on each side; the posterior half of the tail stripe is usually broken into spots. The belly is yellow to reddish orange. Breeding males have a slender cirrus extending downward from each nostril. Southern Two-lined Salamanders are very similar; but their tail stripes usually extend unbroken to the tip of the tail, and they normally have 14 costal grooves, whereas Northern Two-lined Salamanders have 15 or 16. Geographic range is most helpful in separating the two.

Northern Two-lined Salamanders occur from a line extending northeastward from Augusta County in the Shenandoah Valley to Fredericksburg, Virginia, northward to eastern Canada. They live in or near springs, seepages, and streams in hardwood forests and swamps. They occur in Mountain as well as Piedmont streams and often inhabit streams in urban areas.

Courtship occurs in March and April. About a month later, females attach about 12 to 40 eggs to the undersides of rocks and logs, usually in running water. They remain with the eggs until they hatch into aquatic larvae in 4 to 10 weeks. Transformation to the juvenile stage occurs in 1 to 2 years. These salamanders are generalist predators on small aquatic and terrestrial invertebrates.

Chamberlain's Dwarf Salamander
Eurycea chamberlaini

54 to 90 mm (2 to 3.5 in.) This small, slender, recently described species resembles the Dwarf Salamander in having 4 toes on each hind foot but has a light brownish yellow or bronze back and a bright yellow belly. Most individuals have 16 costal grooves, whereas there are usually 18 in the Dwarf Salamander. The under-

surface of the tail is yellow or orange. Southern Two-lined Salamanders are similar but have 5 toes on each hind foot.

Chamberlain's Dwarf Salamanders occur in a variety of habitats but seem to prefer seepage areas near streams or ponds. This species occurs in portions of the Piedmont and in some areas of the Coastal Plain in both Carolinas.

Ovarian egg counts have ranged from 35 to 64. In fall or winter, females deposit eggs, attached singly or in groups of 3 to 8, to the under-surface of leaves or debris in seepages near streams or ponds. Most hatching probably occurs in March. Hatchling larvae have paired dorsal spots, but these disappear in older individuals. Larvae transform in about 2 to 3 months.

Southern Two-lined Salamander
Eurycea cirrigera

64 to 110 mm (2.5 to 4.3 in.) This common salamander is sometimes regarded as a subspecies of *E. bislineata*. It has a yellow, yellowish brown, greenish, or orange dorsal band with black dots or flecks and a dark dorsolateral stripe on each side, usually extending to the tip of the tail. The belly is yellow to orange. There are normally 14 costal grooves. Most breeding males have a slender cirrus extending downward from each nostril and conspicuous mental glands, but some lack cirri and mental glands and have enlarged heads. Sometimes referred to as "morph A," these forms are not well understood but may be related to length of larval period and size at transformation. The salamanders in this species complex differ genetically, but geographic location is most helpful in separating them. Northern Two-lined Salamanders and Blue Ridge Two-lined Salamanders at elevations above about 1,200 m (4,000 ft.) usually have 15 to 16 costal grooves and tail stripes that break into scattered dots along the posterior half. The "Sandhills Eurycea" is more orange or red and lacks distinct dorsolateral stripes. Dwarf and Chamberlain's Dwarf salamanders are similar but have only 4 toes on each hind foot; all members of the Two-lined Salamander complex have 5.

Southern Two-lined Salamanders live in or near springs, seepages, and streams in hardwood forests and swamps. They occur throughout South Carolina, throughout the Piedmont and much of the Coastal Plain of North Carolina, and in most of the southern half of Virginia.

Courtship occurs in fall, and most eggs are laid in winter and spring. The female deposits a flat cluster of eggs on the undersurface of a rock, log, or other object, usually in running water. She remains with the eggs until they hatch into aquatic larvae. The larval stage may last from 1 to 3 years, and the larvae are frequently encountered in small streams.

Three-lined Salamander
Eurycea guttolineata

90 to 200 mm (3.5 to 8 in.) This slender salamander has a black stripe down the back flanked on each side by a pair of tan to yellowish stripes that meet on the base of the tail. The sides are dark brown to black with a light streak between the limbs, and the venter is tan and marbled with greenish gray to black pigment. The vertical bands on the tail may be fused to form a dark, wavy stripe. This species was formerly treated as a subspecies of the Long-tailed Salamander, which it resembles in size and shape but differs from in color pattern and genetic makeup. In our area the two species seldom occur together. Populations in northern Virginia may resemble hybrids with characteristics of both species, but genetic analysis in other parts of the range on similar populations suggests that these are variants of the Three-lined Salamander.

Three-lined Salamanders occur commonly in creek bottomlands in the Piedmont, are locally common in the Coastal Plain, and are restricted to larger valleys in the Mountains.

In late fall and early winter, females move to breeding sites in streams, seepages, and bogs, where they attach their eggs singly in groups of 8 to 14 under rocks or other hard surfaces. Larvae average 10 to 13 mm (0.4 to 0.5 in.) at hatching and usually transform in 4 to 6 months. In some montane populations they may overwinter, requiring 14 to 16 months to transform. Juveniles then live under logs, stones, and debris along stream bottomlands. Sexual maturity is reached the following summer. Adults feed on a variety of small insects and other invertebrates.

Junaluska Salamander
Eurycea junaluska

80 to 100 mm (3 to 4 in.) The dull yellow-brown dorsum is sprinkled with small, dark brown, irregular spots. A wavy, broken black stripe or a series of black blotches extends from the nostril through the eye and along the side of the body. On the tail it becomes a series of small dots or thin lines. The venter is light greenish yellow and lacks mottling. This species has 14 costal grooves. The tail is about half the total length. Males have a well-developed mental gland but lack cirri. The Junaluska Salamander resembles the sympatric Blue Ridge Two-lined Salamander but has longer legs, a shorter tail, and intense mottling on the sides of the body. Two-lined Salamanders also have broader dorsolateral stripes and brighter yellow or orange backs.

Junaluska Salamanders are known only from Graham County, North Carolina, and several counties in adjacent Tennessee, where they live at low elevations under rocks and logs along streams. Adults are most often encountered on roads on rainy nights.

Females lay 31 to 68 eggs in March to May, attaching them singly to the undersides of rocks in streams. They remain with the eggs until they hatch in April to June. Metamorphosing individuals are 70 to 85 mm (2.75 to 3.3 in.) long and transform into juveniles in May to August, after an aquatic larval period of 1 to 2 years. Their diet has not been reported but probably consists of small invertebrates.

The Junaluska Salamander is listed as a Threatened species in North Carolina.

Long-tailed Salamander
Eurycea longicauda

90 to 197 mm (3.5 to 8 in.) This is a slender, yellow to orange-red species with abundant round black spots irregularly spaced on the sides and back. The tail has vertical dark bars on the sides that form a herringbone pattern. The belly is plain yellow to cream. The tail may be nearly two-thirds of the total length in large adults but is proportionately shorter in young animals.

In Virginia this species is found

on the western slopes of the Blue Ridge Mountains and in the Valley and Ridge physiographic province, but in North Carolina it is known only from the Watauga, Nantahala, New, and Little Tennessee river basins. It is usually associated with limestone and shale substrates over most of its range, where it is found along rocky streams and bottom-lands and commonly in damp caves.

The number of eggs ranges from 61 to 106 per clutch, and these are laid underground. In mines and caves the eggs are attached singly to stones or other objects, in or suspended above the water. Larvae hatch in March or April at less than 20 mm (0.75 in.) in length, and transformation usually occurs by June at 40 to 50 mm (1.5 to 2 in.). Sexual maturity is attained the next summer. The diet includes worms and various arthropods.

The Long-tailed Salamander is listed as a species of Special Concern in North Carolina.

Cave Salamander *Eurycea lucifuga*
125 to 181 mm (5 to 7 in.) Cave Salamanders are reddish orange with
round black spots scattered over
the sides and back, sometimes in
irregular rows. They are slender and
flattened and have very long tails.
The young are paler and yellowish.
This species differs from the Long-
tailed Salamander by lacking vertical
dark bars on the sides of the tail and
by having a broader head with more
protrusive eyes.

Cave Salamanders occur from
Rockbridge County, Virginia, south
and west across the Valley and Ridge
physiographic province through the
Cumberland Plateau to the Ozarks.
Virtually restricted to limestone
regions, this species is partial to the
twilight zone of caves and climbs
well on damp walls and ledges, but
it is not a true troglobite. During wet
periods it may occur near springs
and along rocky brooks under logs
and stones. It often occurs in abandoned mine shafts and moist, dark
tunnels.

In September to February, females
deposit about 49 to 87 eggs, attached
to the undersides of rocks in cave
streams or left unattached in small
pools. Hatchlings are sparsely pigmented and only 10 mm (0.4 in.)
long when they appear in January to
February. Metamorphosis occurs at
50 to 60 mm (2 to 2.25 in.), and the
young become sexually mature at
about 125 mm (5 in.) total length 18
months to 2 years later. These salamanders have prehensile tails that
aid in climbing rock faces. They forage inside the twilight zones of caves
and on the outside at night for various insects and other invertebrate
prey.

Dwarf Salamander
Eurycea quadridigitata

54 to 90 mm (2 to 3.5 in.) This small, slender species has a dark brown or bronze back, a dark dorsolateral stripe on each side, a silver-gray belly, narrow whitish streaks along its lower sides, and 4 toes on each hind foot. Breeding males have cirri. This species resembles Chamberlain's Dwarf Salamander but is darker with a silvery belly and differs genetically. Usually the two use different habitats, but they are known to occur together in at least two localities in Allendale and Barnwell counties, South Carolina. Members of the Two-lined Salamander complex may look similar but have 5 toes on each hind foot.

Dwarf Salamanders live beneath surface litter or logs in bottomland forests and swamps and at the edges of pine savanna ponds. In North Carolina they occur primarily in clay-based Carolina bays, cypress savanna ponds, and similar habitats in the southeastern Coastal Plain. In South Carolina they occur in a variety of wetland habitats throughout much of the Coastal Plain.

In late fall or winter, females produce clutches of about 13 to 62 eggs. Nests have not been reported in North Carolina, but in South Carolina the eggs are usually attached singly to moss, leaves, rootlets, or the undersides of logs in seepage areas or near the edges of shallow ponds; they are not attended by the female. Embryonic development requires about 4 to 6 weeks. The aquatic larvae first appear in January and February and transform from about April to June. These salamanders are generalist predators on small invertebrates.

The Dwarf Salamander is listed as a species of Special Concern in North Carolina.

Blue Ridge Two-lined Salamander
Eurycea wilderae
70 to 121 mm (2.75 to 4.75 in.) This salamander is sometimes regarded as a subspecies of *E. bislineata*. Often more brightly colored and patterned than other members of the Two-lined Salamander complex, it has a yellow, yellowish brown, or orange dorsal band with black dots or flecks and a dark dorsolateral stripe on each side, usually extending about midway down the tail, then broken or absent toward the tip. The belly is bright yellow to orange. Populations above 1,200 m (4,000 ft.) usually have 15 to 16 costal grooves; those below those elevations may have 14. Most breeding males have a slender cirrus extending downward from each nostril and conspicuous mental glands, but some lack cirri and mental glands and have broad heads. Similar forms, sometimes referred to as "morph A," also occur in the Southern Two-lined Salamander.

Blue Ridge Two-lined Salamanders live in or near streams, seepages, and springs throughout most of the North Carolina Mountains and in the southern portion of the Blue Ridge province in southwestern Virginia.

The timing of egg laying varies with elevation. Most clutches are probably deposited in late winter and spring, but some nesting may occur in late summer and fall. Females attach clusters of 8 to 56 eggs to the undersurfaces of rocks in streams or seeps and attend them until they hatch into aquatic larvae. The larval stage usually lasts 1 to 2 years.

"Sandhills Eurycea" *Eurycea* n.sp.
60 to 90 mm (2.4 to 3.5 in.) This
member of the Two-lined Salamander species complex has long been
recognized as distinct but has yet to
be formally described. It is similar to
the Southern Two-lined Salamander
but lacks the distinct dorsolateral
dark stripes characteristic of that
species. The dorsal body color is red
to burnt orange with black dots or
flecks. The belly is reddish orange.

Mature males tend to be darker and
more orange; females, more red.
Breeding males have a slender cirrus extending downward from each
nostril. The larvae are very similar
to those of the Southern Two-lined
Salamander. Chamberlain's Dwarf
Salamanders and Dwarf Salamanders are similar but more slender
and have only 4 toes on each hind
foot; the "Sandhills Eurycea" has 5.

The "Sandhills Eurycea" lives in
or near springs and streams with
cool, flowing water. It is apparently
endemic to the Sandhills region of
North Carolina. Range overlap with
the Southern Two-lined Salamander
has been found at only one site.

Eggs are laid in the winter and
spring. The female deposits a cluster
of eggs among root masses in the
edge of a stream. Limited observations suggest she remains with the
eggs until they hatch into aquatic
larvae.

Spring Salamander
Gyrinophilus porphyriticus

120 to 220 mm (4.5 to 8.5 in.) This rather large salamander has a stout body, a broadly truncate snout, and 18 costal grooves. The dorsum is light brownish orange, red, or salmon, often with small dark spots or flecks. A light line, bordered below by a dark line, extends from eye to nostril. The venter is flesh-colored, and the throat may be flecked or reticulated with black. Spring Salamanders resemble Mud and Red Salamanders but are more agile and have broader and flatter snouts.

Spring Salamanders inhabit springs and other small, cold, rocky streams, and caves in the Mountains and the upper Piedmont of Virginia and the Carolinas. They hide under stones near the edges of streams during the day but forage for prey at night. They are active in water year-round.

Usually in July or August, a female attaches 16 to 106 eggs to the lower surface of a submerged rock or other object and attends them until they hatch in about 3 months. Hatchlings average 18 to 22 mm (0.7 to 0.8 in.) total length. The larval stage may last up to 4 years. Larvae have a purplish ground color. These are relatively long-lived salamanders. Females are about 5 years old at first reproduction. Spring Salamanders are notorious for preying on smaller salamanders, but invertebrates such as insects, worms, slugs, and snails are also eaten.

Four-toed Salamander
Hemidactylium scutatum

50 to 95 mm (2 to 3.5 in.) This small brown salamander has a white belly with conspicuous scattered black dots; a reddish tail with a noticeable constriction at the base, where it breaks easily; and 4 toes on each hind foot.

Although this species is widely distributed in the northeastern United States, its populations are scattered in our region. This bog dweller requires seepages or shallow ponds with moss-covered logs, roots, and grass clumps over quiet

water. The Four-toed Salamander is known from only a few localities in the Coastal Plain of the Carolinas. It is more widespread in the Piedmont and Mountains and occurs throughout most of Virginia.

Females lay about 30 to 80 eggs, usually under clumps of moss in woodland pools or seeps, and attend them until hatching. Large aggregations of eggs and females may occur together. In our area, egg laying takes place from mid-February to May, and hatching occurs in May to June. The larvae, about 11 to 15 mm (0.5 to 0.6 in.) long at hatching, wriggle down into the water, where they remain and grow for about 6 to 8 weeks. At about 18 to 24 mm (0.7 to 1 in.) total length they emerge as terrestrial juveniles and mature 1.5 years later. Larvae feed on zooplankton and small invertebrates, and adults are known to eat arthropods and other small invertebrates.

The Four-toed Salamander is listed as a species of Special Concern in North Carolina.

Blue Ridge Gray-cheeked
Salamander *Plethodon amplus*

Cheoah Bald Salamander
Plethodon cheoah

Jordan's Salamander
Plethodon jordani

South Mountain Gray-cheeked
Salamander *Plethodon meridianus*

Southern Gray-cheeked
Salamander *Plethodon metcalfi*

Northern Gray-cheeked
Salamander *Plethodon montanus*

Red-legged Salamander
Plethodon shermani

90 to 184 mm (3.5 to 7.25 in.) The
species in this large woodland salamander group have been variously
treated as species, as subspecies,
and as ecological variants. Included
here are those recently recognized
as distinct species in our area. Most
forms are basically plain black or
gray and resemble unspotted members of the Slimy Salamander complex. Various members of these two
complexes occur together without
hybridizing in some areas; in others
they produce hybrid swarms.

Plethodon jordani occurs in the
Great Smoky Mountains of North
Carolina and Tennessee and is
easily identified by its bright red
cheek patches. *Plethodon shermani*
occurs in the Nantahala and Tusqui-
tee mountains and has bright red
patches on the dorsal surfaces of its
legs. *Plethodon cheoah* is endemic to
Cheoah Bald in Graham and Swain
counties, North Carolina, and also
has red leg patches. The remaining
"gray-cheeked" species normally are
plain gray or black without bright leg
or cheek patches. *Plethodon meridianus* is endemic to the South Mountains of Burke, Rutherford, and
Cleveland counties, North Carolina.
Plethodon amplus occurs in the Hickory Nut Gorge vicinity in portions
of Buncombe, Henderson, McDow-
ell, Polk, and Rutherford counties,
North Carolina. *Plethodon montanus*
occurs in southern portions of the
Blue Ridge and Valley and Ridge
physiographic provinces in western
Virginia, in the northern portion of
the Blue Ridge Mountains in North
Carolina, and in adjacent Tennessee. *Plethodon metcalfi* occupies the
southern Blue Ridge Mountains of
southwestern North Carolina and
northwestern South Carolina, west
of the French Broad River. Some
individual *P. metcalfi* from extreme
northwestern South Carolina and
southern Jackson and Macon counties, North Carolina, are marked

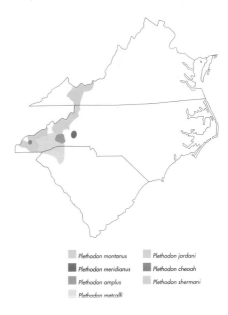

Plethodon montanus   Plethodon jordani
Plethodon meridianus   Plethodon cheoah
Plethodon amplus   Plethodon shermani
Plethodon metcalfi

Blue Ridge Gray-cheeked Salamander

Cheoah Bald Salamander

Jordan's Salamander

with conspicuous gold, silver, or brassy frosting and were formerly referred to as *P. clemsonae* ("Clemson Salamander").

Some Imitator and Ocoee Salamanders closely resemble *P. jordani*, *P. shermani*, and *P. cheoah* by having red cheeks, red legs, and dark adult color, but these possible Batesian mimics can be recognized by their head shape, the light bar between the eye and the jaw, and their larger hind legs.

The species in this complex are usually found at elevations of 600 m

South Mountain Gray-cheeked Salamander

Southern Gray-cheeked Salamander ("Clemson phase")

Northern Gray-cheeked Salamander

Red-legged Salamander

(1,970 ft.) or higher, with a few exceptions. They inhabit the forest floor and are active on the surface at night when the humidity is high and temperatures are mild. During the day, they occupy root tunnels or burrows under stones and logs. All salamanders in the genus *Plethodon* breed terrestrially. Larval development takes place inside the eggs, and the terrestrial hatchlings resemble miniature adults. The reproductive habits of the Jordan's Salamander complex are poorly known, despite their frequent abundance. Nests have not been reported, but eggs are presumably deposited in underground chambers in late spring or early summer and hatch in late summer or fall.

Tellico Salamander
Plethodon aureolus

100 to 151 mm (4 to 6 in.) This recently described species closely resembles members of the Slimy Salamander complex, but it is smaller and differs genetically. It has a grayish or black dorsum with abundant brassy spots, a light-colored chin, a concentration of white or yellowish spots along its lower sides, and usually 16 costal grooves. Its brassy

spots should distinguish it from other woodland salamanders within its small range; Southern Appalachian Salamanders and members of the Slimy Salamander complex have white spots.

The Tellico Salamander is known only from the western slopes of the Unicoi Mountains and nearby lowlands in Cherokee and Graham counties, North Carolina, and adjacent Monroe and Polk counties, Tennessee. It occurs mostly at lower elevations but is also known from mountainous habitats up to 1,620 m (5,320 ft.). It is most often found under logs and other sheltering objects in forested areas.

Reproduction is terrestrial, as in other members of the genus. Juveniles have been found in August, but most aspects of this salamander's life history and ecology are unknown.

SLIMY SALAMANDER COMPLEX

Chattahoochee Slimy Salamander
Plethodon chattahoochee

Atlantic Coast Slimy Salamander
Plethodon chlorobryonis

White-spotted Slimy Salamander
Plethodon cylindraceus

Northern Slimy Salamander
Plethodon glutinosus

South Carolina Slimy Salamander
Plethodon variolatus

120 to 206 mm (4.5 to 8 in.) The salamanders in this confusing group were once treated as a single species (*P. glutinosus*), but recent genetic analyses have shown them to be a complex of closely related species. Some authorities advocate recognizing them as geographic variants rather than separate species. They are not easily or reliably distinguished in the field and, aside from laboratory techniques, are best identified by range. All are black with white or cream flecks scattered over the sides and sometimes on the back and limbs. Glands in the skin, especially on the tail, exude a sticky slime that is difficult to remove from the hands.

Most wooded habitats in our region harbor at least one member of this complex. Exceptions include bottomlands subject to repeated flooding, some of the higher mountain peaks, most barrier islands, the Delmarva Peninsula, and portions of the southwestern Mountains, where they are replaced by *P. teyahalee*.

Plethodon cylindraceus is most widespread in our area, ranging through most of the Piedmont and Mountains of Virginia, North Carolina, and northwestern South Carolina. It is usually fairly heavily spotted with white.

Plethodon chlorobryonis ranges from the southeastern Coastal Plain of Virginia through most of the Coastal Plain of the Carolinas, west into portions of the South Carolina Piedmont. It is usually relatively small and the least heavily patterned species, often with few or no dorsal white spots but sometimes with heavy white or yellowish "frosting" concentrated along its lower sides.

Plethodon glutinosus occurs in western Virginia and probably enters extreme northwestern North Carolina. Its white spotting is often extensive.

Plethodon variolatus occurs in the southeastern Coastal Plain of South Carolina and is highly variable in pattern.

Plethodon chattahoochee barely enters our area, probably only in Cherokee and Clay counties, North Carolina. It may have heavy concen-

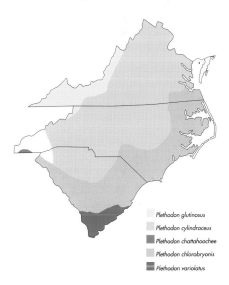

Plethodon glutinosus
Plethodon cylindraceus
Plethodon chattahoochee
Plethodon chlorobryonis
Plethodon variolatus

Atlantic Coast Slimy Salamander

Northern Slimy Salamander

White-spotted Slimy Salamander

South Carolina Slimy Salamander

Chattahoochee Slimy Salamander

trations of whitish pigment along its lower sides and often only small, scattered dorsal spots.

Slimy Salamanders are active near the surface year-round but usually burrow deep during hot, dry periods and hard freezes. They spend the day in burrows under logs, stones, and leaf litter, foraging at night for arthropods, worms,

and other invertebrates. Mating and nesting times vary with locality. Eggs are laid in clusters of about 5 to 38, suspended from the ceiling of a natural cavity, usually underground. Females usually attend their nests, which are rarely found. The terrestrial juveniles hatch in about 2 to 3 months.

Red-backed Salamander
Plethodon cinereus

57 to 127 mm (2.25 to 5 in.) This small salamander usually has a straight-edged, reddish dorsal stripe and a heavily mottled black and white belly. Unstriped, uniformly dark gray or black individuals ("lead-backs") with scattered brassy or white (occasionally reddish) flecks occur in many populations. Reddish versions of this salamander have been found in northern Virginia. Southern Zigzag Salamanders have a dorsal stripe with zigzag edges, at least anteriorly, and reddish orange pigment over much of the belly. Southern Ravine Salamanders and Peaks of Otter Salamanders are more slender with longer tails and distinctly darker bellies. Shenandoah Salamanders have narrower reddish stripes on the back and dark bellies with only a few light markings. Big Levels Salamanders are very similar but have larger heads and only occur in the Big Levels area in Virginia.

Red-backed Salamanders live under rocks, leaf litter, and decomposing logs in mixed pine and hardwood forests over a wide range in northeastern North America. They occur throughout much of Virginia and in the Mountains north and east of the French Broad River, as well as in portions of the Piedmont and Coastal Plain, in North Carolina.

The mating period extends from October to April except in the coldest months. In May or June, a female deposits from 1 to 14 eggs, usually within a cavity underground, under rocks, or in a decomposing log. She attends the eggs until they hatch into terrestrial young in late August or September. These salamanders are generalist predators on small invertebrates.

Valley and Ridge Salamander
Plethodon hoffmani

80 to 137 mm (3 to 5.5 in.) This slender salamander has short legs, a long tail, and usually 21 costal grooves. The dorsum is dark brown to blackish with scattered whitish or brassy flecks. The venter is dark with white mottling, especially on the chin. Most specimens from along the West Virginia border have but 20 costal grooves, and some have a narrow dorsal red stripe. *P. hoffmani* has 1 or 2 more costal grooves than the similarly proportioned *P. hubrichti* and *P. richmondi*. In addition, the chin and belly are lighter in *P. hubrichti* but are darker in *P. richmondi*.

This salamander lives under logs and rocks in the Valley and Ridge physiographic province north of the New River in Virginia to Rockingham County. These salamanders tolerate cool weather well; they appear early in March and disappear in late summer when the surface is dry. They are again active during the early half of fall.

Females lay 3 to 8 eggs in underground cavities in late May to early June and attend their eggs until hatching in late August to September. The young mature in 2 years. Like most terrestrial salamanders, they eat many different kinds of insects as well as worms and other invertebrates.

Peaks of Otter Salamander
Plethodon hubrichti

80 to 122 mm (3 to 5 in.) The dorsum is black with numerous brassy flecks often forming small spots or blotches. A narrow reddish dorsal stripe is found only in hatchlings, although some older individuals may have a roughly continuous coppery stripe. The venter is black or dark slate gray. A few small white spots occur on the cheeks and sides of the body. This species usually has 19 costal grooves. As in most *Plethodon*, the tail is slender and nearly circular in cross section. Males have a large mental gland during the breeding season. The abundant brassy flecks on the dorsum and lack of confusing species in its range make identification easy.

The Peaks of Otter Salamander occurs only in cool, moist hardwoods in the Blue Ridge Mountains in Bedford, Botetourt, and Rockbridge counties, Virginia, from Flattop Mountain at the Peaks of Otter to a point 19 km (12 mi.) north along the Blue Ridge Parkway. Individuals are active on the surface during wet periods from about March through October.

In May or June, females lay clusters of 8 to 10 eggs in or under decaying logs and rocks embedded in the ground. Hatchlings appear in late August through September. During warm, wet nights these terrestrial salamanders forage for a wide variety of invertebrates on the ground, on tree trunks, and on herbaceous vegetation.

Cumberland Plateau Salamander
Plethodon kentucki
98 to 168 mm (3.8 to 6.6 in.) The Cumberland Plateau Salamander is black with small, widely scattered white spots that are more concentrated on the sides, with some brassy flecking on some individuals. The venter is dark gray, but the chin is almost cream. The eyes of this species protrude outward slightly more than those in the sympatric Northern Slimy Salamander, which also has a dark chin and much larger

and more abundant white dorsal spots. Wehrle's Salamanders in the range of the Cumberland Plateau Salamander have dark bodies with yellow dots.

In our area, this species occurs only in the far southwestern portion of Virginia. It ranges into eastern Kentucky, southern West Virginia, and a few counties in Tennessee, where it is most often found under logs and rocks in moist mixed hardwoods. The activity period is usually April through October except during dry periods, when they move underground.

Mating occurs on land. Females subsequently produce 9 to 12 eggs, which they lay in cavities in the ground during July and August. Hatchlings appear on the surface in September and October. These salamanders forage at night for a wide range of insects and other invertebrates.

Cow Knob Salamander
Plethodon punctatus

100 to 157 mm (4 to 6 in.) This blackish salamander has many small white or yellowish white spots on the dorsum and a few brassy flecks on the head and tail. The throat is light, but the remainder of the venter is uniformly dark gray to black. There are 17 or 18 costal grooves. This species is similar to Wehrle's Salamander but has more webbing between the toes and different dorsal markings: larger and more numerous white spots, fewer brassy flecks, and no red spots on juveniles. Northern Slimy Salamanders in this region are black with silvery white spots on the back and sides.

The range is restricted to Shenandoah Mountain in Virginia and West Virginia and southward along the ridge line into Augusta County, Virginia. This species is restricted to elevations above 810 m (2,660 ft.) in mature hardwood forests with abundant rock cover.

By day, Cow Knob Salamanders hide under rocks and logs or in burrows, but at night they forage in the open for insects, earthworms, snails, spiders, and other invertebrates. They emerge from hibernation in April, are active on moist days in spring through fall, and retreat into deep burrows by late October. Females deposit 7 to 16 eggs beneath rocks or in underground chambers. Hatchlings appear in September and mature in 3 years.

Southern Ravine Salamander
Plethodon richmondi
76 to 143 mm (3 to 5.5 in.) This long, slender salamander lacks conspicuous markings. The small legs are far apart, separated by 19 to 22 costal grooves. The dorsum is dark brown or almost black with scattered silvery or brassy flecks. The venter is uniformly dark except for some light blotches on the throat and lower sides of the body. Little individual and geographic variation occurs; however, some specimens have red pigment on the cheeks, front legs, and anterior sides of the body. A narrow red dorsal stripe exists only in late embryos. This species differs from the other small woodland salamanders by having more costal grooves and a darker belly and throat.

These salamanders inhabit southwestern Virginia south of the New River and northwestern North Carolina. They prefer high, moist hardwoods. Individuals are hard to find during dry summers, but in spring and fall large numbers occur under thin, flat rocks and other surface cover.

Mating occurs in April and June, and females lay 5 to 11 eggs underground about a month later. Hatching occurs in late August to September. The terrestrial hatchlings are about 25 mm (1 in.) long. Like other woodland salamanders, this species forages on the ground and on vegetation for a wide variety of insects and other small invertebrates.

Southern Red-backed Salamander
Plethodon serratus

57 to 127 mm (2.25 to 5 in.) This cryptic species is very similar to the Red-backed Salamander. The two were long regarded as a single species, but they differ genetically. Few external differences exist, and those that do are not truly diagnostic. Unstriped or "leadback" *P. serratus* are rare and virtually always have at least some dorsal red pigment; unstriped *P. cinereus* usually do not. Red pigment is usually present on the anterior part of the belly in *P. serratus* but not in *P. cinereus*, and the red dorsal stripe is often more extensive on the tail in *P. serratus*.

Both species of Red-backed Salamanders live in similar habitats, but in our area they are geographically isolated. The Southern Red-backed Salamander occurs in the North Carolina Mountains south and west of the French Broad River. It has not been reported in South Carolina. This is a cool-weather species, moving into deeper underground retreats between April and October when it is too warm for surface activity.

Mating occurs in winter. In June and July, females deposit about 5 to 7 eggs in deep underground chambers. Hatching occurs two months later. These salamanders forage on the forest floor for small insects and other invertebrates.

Shenandoah Salamander
Plethodon shenandoah

76 to 110 mm (3 to 4.5 in.) This small, dark salamander has two color phases. Some individuals have a narrow red or yellow stripe on the back, and others have only scattered brassy flecks and small red spots. Striped individuals constitute about 50 percent of southern populations and 100 percent of northern ones. The venter is black with only a few small white or yellow spots. Costal grooves average 18 per individual. Red-backed Salamanders are very similar but have an average of 19 costal grooves, a broader and darker dorsal stripe, a lighter and more mottled venter, and a more narrow head.

Shenandoah Salamanders occur above 900 m (2,950 ft.) only on three high mountains in the Shenandoah National Park in northern Virginia. They are surprisingly tolerant of dry conditions found on steep, north-facing talus slopes, but they are restricted to that habitat because of competition with Red-backed Salamanders that occur in the surrounding forest.

These salamanders are active from April to October but retreat underground when the surface is too dry. Females lay 4 to 19 eggs in moist crevices underground. Parental care is unknown. Young of the year emerge in September. They forage on warm, rainy nights on the surface and by climbing vegetation. Prey include worms, millipedes, centipedes, spiders, and a variety of insects.

The Shenandoah Salamander is federally listed as an Endangered species.

Big Levels Salamander
Plethodon sherando

91 to 112 mm (3.8 to 4.4 in.) This recently described salamander is very similar to the Red-backed Salamander. A reddish stripe occurs from the head down the back onto the tail. Individuals without stripes are dark gray to nearly black with tiny brassy flecks. The belly has a salt-and-pepper pattern in both forms. The head is about 25 percent wider in Big Levels Salamanders than in Red-backed Salamanders the same size. Big Levels Salamanders have longer limbs, and the number of costal grooves between the limbs when they are folded along the body is 1 to 5, compared with 5.5 to 9 in Red-backed Salamanders.

Big Levels Salamanders occur only in the Big Levels area of Augusta and Nelson counties in Virginia. Their life history is very similar to that of the Red-backed Salamander. They are most abundant in mixed hardwood forests, where they live under surface objects and in the ground among the interstices in the soil.

Little is known about reproduction in this species. Females probably lay 3 to 8 eggs in underground crevices in spring. The eggs hatch in late summer and fall. Adults and juveniles eat a diverse array of insects and other invertebrates.

Southern Appalachian Salamander
Plethodon teyahalee (or Plethodon oconaluftee)

121 to 207 mm (4.75 to 8.2 in.) This large woodland salamander is gray or black with very small white dorsal spots, often larger white spots along the sides, and sometimes small red spots on the legs. The belly is uniformly slate gray; the chin and throat are paler. There are usually 16 costal grooves. This species closely resembles members of the Slimy Salamander complex, which it replaces in most of its range. Some hybridization may occur where ranges overlap. It also hybridizes extensively with some members of the Jordan's Salamander complex. Some authors refer to this species as *P. oconaluftee*. Its nomenclatural status has been debated, largely because the holotype (the specimen upon which the original species description was based) was from a hybrid zone.

This species occurs in the southwestern Mountains of North Carolina west of the French Broad River and in Abbeville, Anderson, Oconee, and Pickens counties, South Carolina, as well as in small portions of adjacent Georgia and Tennessee. It inhabits forested areas, where it is most often found beneath logs or rocks or foraging on the forest floor at night.

Little is known about the life history of this species. Nests have not been reported, but females apparently deposit eggs underground in late spring or summer, and hatchlings appear in late summer and early fall. Insects and millipedes are frequent items in the diet, which also includes a variety of other invertebrates.

Southern Zigzag Salamander
Plethodon ventralis

64 to 111 mm (2.5 to 4.5 in.) This small salamander has a reddish orange to reddish brown stripe down the brown back and tail with wavy or zigzag edges at least on its anterior half. The belly is heavily mottled with black, white, and reddish orange. Many individuals in our area have no stripe, and both color phases have a variable amount of red pigment on the back, sides, and belly. Red-backed Salamanders have a straight-edged reddish dorsal stripe and a salt-and-pepper belly (if ventral reddish orange pigment is present, it is confined to the area between the front limbs).

Southern Zigzag Salamanders are usually associated with seepages near rock outcrops on slopes in mixed hardwood forests and live under leaf litter, rocks, or decomposing logs. They may be found in caves in summer months. This rare species is known in our area only in Buncombe, Henderson, and Madison counties, North Carolina, and Scott and Washington counties, Virginia.

It also occurs in southeastern Kentucky, eastern Tennessee, and northwestern Georgia.

Little is known about the life history of this salamander in our area. Most surface activity occurs in spring, fall, and relatively warm periods in winter. Individuals become increasingly difficult to find at other times. Females deposit about 3 to 10 eggs in underground cavities in spring, and hatchlings appear in late summer. Their diet has not been reported.

The Southern Zigzag Salamander is listed as a species of Special Concern in North Carolina.

Shenandoah Mountain Salamander *Plethodon virginia* 91 to 112 mm (3.8 to 4.4 in.) This recently described salamander is difficult to identify except by location. It closely resembles the Valley and Ridge Salamander. Adults are brownish black with many small white spots and brassy flecks, usually most prominent on the sides. In some areas, individuals have narrow reddish dorsal stripes. The belly and throat are black with some white mottling. The "leadback" phase of

the Red-backed Salamander closely resembles this species but has a salt-and-pepper belly.

Shenandoah Mountain Salamanders occur from Great North Mountain south along the higher elevations of Shenandoah Mountain down into Augusta County, Virginia. These woodland salamanders remain underground in small spaces between the rocks during cold or dry weather but emerge on rainy nights to forage from April through September. They are very difficult to find in winter months. They prefer mature hardwood forests, although they will inhabit other forest types.

Females lay 3 to 8 eggs in underground crevices in May and June. Hatching occurs in August and September. Juveniles emerge from their underground cavities the following spring. Adults and juveniles eat a wide array of prey, including worms, snails, spiders, millipedes, and a variety of insects.

Webster's Salamander
Plethodon websteri

60 to 85 mm (2.5 to 3.3 in.) This small species closely resembles the Southern Zigzag Salamander. Both are brown with or without a wavy yellowish brown to reddish stripe on the back. Individuals lacking the dorsal stripe usually have scattered red pigment. Both forms have tiny silvery white spots and brassy flecks that yield a frosty appearance. The dorsum varies from bright to dusky reddish orange and is brightest on the tail, which typically has less dark pigment than the body. The venter is mottled with black, white, and reddish orange, with the orange especially prominent on the neck. Individuals from our area differ from those in other parts of the range by having a mode of 19 rather than 18 costal grooves.

Webster's Salamander is known to occur from Edgefield, Greenwood, and McCormick counties, South Carolina, westward into Mississippi, generally south of the range of the Zigzag Salamander. It inhabits moist, mixed hardwood forests on steep, north-facing slopes with rock outcrops, where it may be found under rocks and logs from spring to fall when the ground is moist. Individuals move into underground retreats during warm, dry months.

In South Carolina, females deposit 3 to 8 eggs underground in June or July and attend them until they hatch in August or September. Adults and young emerge and are active on the forest floor from October until May or June. Individuals remain underground in the summer.

Webster's Salamander is listed as an Endangered species in South Carolina.

Wehrle's Salamander
Plethodon wehrlei

100 to 160 mm (4 to 6.5 in.) This species usually is dark gray or brown with an irregular row of white, bluish white, or yellow markings on the sides of the body. The venter is uniformly gray except for white blotches on the throat. Juveniles have reddish dorsal spots, arranged roughly in pairs. Cave populations are highly variable in color and pattern; for example, individuals from caves near Roanoke, Virginia, are purplish brown with numerous light flecks and bronze mottling. Costal grooves range from 16 to 18 and average 17.

This species resembles members of the Slimy Salamander complex but is more slender and has more costal grooves.

Wehrle's Salamanders occupy upland forests in western Virginia and a few localities in Alleghany, Stokes, Surry, and Wilkes counties, North Carolina. They inhabit the entrances of caves and deep rock crevices, as well as the forest floor, where they may be found under rocks and logs. They may be active year-round, but surface activity peaks in spring and fall.

Females deposit 9 to 24 eggs in late spring and early summer in damp logs, soil, or moss and in crevices in caves. They remain with the eggs until hatching occurs in September. Juveniles and adults forage on the forest floor and climb vegetation or rock faces during wet periods to catch a wide variety of insects and other invertebrates.

Wehrle's Salamander is listed as a Threatened species in North Carolina.

Weller's Salamander
Plethodon welleri

64 to 79 mm (2.5 to 3 in.) This small black salamander is profusely flecked with brass or gold. The flecks are often fused, forming large, irregular blotches or spots. The belly is usually mottled with white but is uniformly dark on specimens from Grandfather Mountain, North Carolina. This species was named after W. H. Weller, a young Ohio naturalist who died in a fall while collecting salamanders on Grandfather Mountain.

Weller's Salamanders occur chiefly in spruce-fir forests on high mountains above 1,500 m (4,900 ft.). During the day, they hide under rocks, leaf litter, and decomposing logs. Populations tend to be associated with talus slopes or other rocky substrates. The small range of this species includes Mount Rogers and Whitetop Mountain in southwestern Virginia and several mountains in northwestern North Carolina and northeastern Tennessee.

Courtship and mating occur in spring and fall. Females deposit small clusters of 4 to 11 eggs beneath mats of moss on decomposing conifer logs, probably in early or mid-summer. Females remain with their eggs until they hatch and the young disperse in late August or early September. The young are miniature replicas of the adults. Known prey include spiders, mites, snails, and numerous insects.

Weller's Salamander is listed as a species of Special Concern in North Carolina.

Yonahlossee Salamander

Yonahlossee Salamander
Plethodon yonahlossee
110 to 190 mm (4.5 to 7.5 in.) This large, handsome salamander is readily recognized in life by the brick red back and profuse gray or white spotting on the lower sides. The red colors fade rapidly in preserved specimens, which then resemble Slimy Salamanders except for their lighter, more mottled throats. Individuals from the Bat Cave–Chimney Rock vicinity of North Carolina have longer legs and more flattened

heads, and the red dorsal color is reduced to small blotches or flecks. These populations usually inhabit rock crevices and were originally described as a separate species— *P. longricus*, the Crevice Salamander. They are, however, very similar genetically and are currently treated as an ecological variant of *P. yonahlossee.*

Yonahlossee Salamanders occur from mountain valleys to 1,700 m (5,575 ft.) on the mountains of the southern Blue Ridge of Virginia through extreme northeastern Tennessee and western North Carolina east of the French Broad River Valley. They inhabit hillsides and ravines, often where rock slides are thickly carpeted with mosses and ferns. Though locally common, they are more restricted in habitat than most large woodland salamanders. They use long burrows in the forest floor and often sit at the entrances waiting for prey or for nightfall, when they emerge to forage.

"Crevice Salamander"

Females attach their 19 to 27 eggs to rocks in small underground cavities in late spring and early summer. Hatchlings appear on the surface in September through October. The diet includes insects and other arthropods, worms, and snails. The common and scientific names, of Native American origin, derive from Old Yonahlossee Road northeast of Linville, North Carolina, whence this species was first described.

The "Crevice Salamander"—that is, the Hickory Nut Gorge population—is protected as a taxon of Special Concern in North Carolina.

Mud Salamander
Pseudotriton montanus

73 to 195 mm (3 to 7.5 in.) This robust salamander has a short tail, brown eyes, and 17 costal grooves. The dorsum is coral pink, bright red, brownish, or light yellowish orange; the lower sides are red or yellow. The dorsal spots are small, distinctly round, black, and well separated. Older adults are darker; the dorsum is reddish to brown with obscure spots, and the venter is flecked or mottled with brown. Red Salamanders have yellow eyes and longer snouts.

Despite their species name, Mud

Salamanders are primarily lowland animals, inhabiting the Coastal Plain and Piedmont and some lowlands in the Mountains. They occur in the fine, black muck beneath logs and stones or in burrows along the banks of seepages, springs, brooks, or swamps at elevations below 700 m (2,300 ft.).

Courtship occurs in early fall. Eggs are usually laid in December and hatch into aquatic larvae from January to March. Ovarian egg counts have ranged from 66 to 192, but natural nests of more than 30 eggs have not been reported. Among the few nests that have been found, eggs have been attached to leaves or rootlets in springs and seeps. Females may lay eggs only every other year. Most larvae transform in about 17 months, but some require an additional year. Average snout-vent length is 10 mm (0.4 in.) at hatching and 36 mm (1.4 in.) at metamorphosis. This species is known to eat a variety of invertebrates as well as smaller salamanders.

Red Salamander
Pseudotriton ruber

76 to 180 mm (3 to 7 in.) This handsome salamander has a short tail and usually 16 costal grooves. The dorsum of a young adult is bright red or orange with many irregular black spots, and the belly is salmon red with dark dots. A row of black flecks may border the mouth. Older animals are dark orange or purplish brown with enlarged, fused dorsal spots and more ventral ones. Yellow or golden eyes and a longer snout distinguish the Red from the Mud Salamander.

This species inhabits most of Virginia, except the extreme southeastern corner and the Eastern Shore, and most of the Carolinas, except for the majority of the Coastal Plain, where it is known mostly from the Sandhills and from several counties along the Savannah River. Adults live in leaf accumulations in spring-fed brooks and nearby crevices and burrows. They also wander far from wetlands when the forest floor is moist and live under logs, boards, stones, and leaves in more terrestrial habitats.

Courtship occurs in summer, egg laying in October, and hatching in early December. Clutches of 29 to 130 eggs are attached to the undersides of rocks and other objects in water. Newly hatched larvae have been found from November to March. The larval period lasts about 32 months, and the average newly metamorphosed salamander is 70 mm (2.75 in.) long. Earthworms, insects, slugs, snails, spiders, millipedes, and small salamanders are included in the diet.

Many-lined Salamander
Stereochilus marginatus

64 to 114 mm (2.5 to 4.5 in.) This yellowish brown salamander has light and dark longitudinal streaks or lines on its lower sides. The usually unmarked back sometimes bears a few small, indistinct dark or light spots. The belly is pale dusky yellow with scattered brown flecks. The head is small, narrow, and flattened. The relatively short tail is keeled and laterally compressed.

Many-lined Salamanders inhabit swamps, ditches, sluggish streams, and shallow cypress or gum ponds in pine savannas. They are almost entirely aquatic but in dry weather hide under leaf litter, sphagnum mats, or moist, decomposing logs. The range of this species in our area extends from southeastern Virginia through the Coastal Plain of the Carolinas.

In January and February, females attach clusters of 6 to 121 eggs to the stems of aquatic mosses, twigs, or the undersurfaces of submerged logs, often 10 to 15 cm (4 to 6 in.) underwater. Females usually remain with their eggs until they hatch in March or April. The aquatic larvae hide in bottom debris and sphagnum mats and transform in late spring and summer, typically after about 2 years. Larvae and adults feed in water, where they consume amphipods, isopods, small clams, and aquatic insects.

Order Anura

FROGS AND TOADS

This large and widely distributed group includes about 2,300 species. Some are aquatic; others, arboreal. Many are highly terrestrial, and a few even inhabit regions generally thought of as too harsh for amphibians: deserts, brackish water, and the Arctic. Adults have many unique anatomical features: a stocky body with an extremely short backbone (only 9 or 10 vertebrae), a broad head usually with large eyes, no tail, very long hip bones, and long hind legs. In our area, adults range in size from 13 mm (0.5 in.) (Little Grass Frog) to 200 mm (8 in.) (American Bullfrog). There are no concise differences between frogs and toads. In general, the term "toad" pertains to the shorter-legged species and customarily includes the bufonids, pelobatids, and microhylids (see species list).

Other important features of most anurans include lungs, keen hearing, and a well-developed vocal apparatus. Vocalization is especially important during the breeding season when myriads of individuals assemble in rain puddles and ponds. There, the persistent calls of the male play a major role in bringing the sexes of each species together. Hedonic glands are absent. The male clasps the female's groin or axillary region and fertilizes the eggs as they are extruded from her cloaca. All of our species lay their eggs in freshwater. The eggs develop into bizarre larvae popularly known as tadpoles. Tadpoles have many distinct features: large head-trunk region, long muscular tail, cryptic gills, keratinized mouthparts, and no true teeth. Metamorphosis is obligatory and marked by drastic morphological changes. Overt paedomorphosis (retention of larval features), as sometimes occurs in the Caudata, is lacking. All our species are carnivorous as adults, feeding on insects and other small animals.

Of the 33 recognized anuran species known from the Carolinas and Virginia, most are hylids: 8 *Hyla*, 8 *Pseudacris*, and 2 *Acris*. The only other large groups are the ranids (true frogs) with 9 species, all in the genus *Rana* (= *Lithobates*) and the bufonids with 4 species of *Bufo* (= *Anaxyrus*). *Hyla*, *Rana*, and *Bufo* are nearly cosmopolitan genera, but all of our other genera are restricted to North America. None of our species is endemic, but 3 have over two-thirds of their geographic ranges in the Carolinas and Virginia. In contrast to the Caudata, which include many montane species, the Anura predominantly inhabit the Coastal Plain, especially in the Carolinas.

the lower belly is light gray or reddish. A pair of large pectoral glands opens onto the chest. The large, protruding eye has a vertical, elliptical pupil. A keratinized, black, spadelike projection is located on the inner border of the hind foot. This effective digging tool enables a squatting toad to burrow in loose soil while making only a slight rocking motion with the hind legs. During dry seasons, Spadefoots may remain torpid underground for months. They can survive enormous water loss (over 40 percent of body weight).

Eastern Spadefoot
Scaphiopus holbrookii

45 to 72 mm (2 to 3 in.) This soft-bodied anuran has a distinct tympanum, a small round parotoid gland, and short legs that are often held close to the body. Small tubercles are scattered over the moist skin. The dorsum is brown, gray, olive, or black. A light band extends over the back from eye to vent, and another extends along the side of the body. These lyre-shaped bands are more yellowish in males than in females. The throat and chest are white, and

Being subterranean and nocturnal, Spadefoots are cryptic animals, but they may be found in abundance when rainfall is extensive and are occasionally encountered foraging on rainless nights. They prefer sandy lowlands and are abundant in parts of the Coastal Plain and lower Piedmont, absent in most of the upper Piedmont, and scattered in the Mountains. Insects and other invertebrates are the chief foods. Spadefoots produce a musty, peppery skin secretion irritating to mucous membranes and evoking an allergic reaction in some people; wash hands thoroughly after handling.

Spadefoots breed in shallow, temporary pools formed after heavy rains. They may breed at almost any time of year but often skip years when the weather is unsuitable. The call is a short, explosive, low-pitched "wank" repeated every 2 seconds. Eggs and tadpoles develop rapidly, requiring 20 to 30 days to transform.

American Toad
*Bufo americanus (*or *Anaxyrus americanus)*

50 to 107 mm (2 to 4.25 in.) This large toad has a short, broad body and a broadly circular snout. The largest specimens inhabit the Mountains. The dorsum may be brown, gray, olive, or red. Although some individuals are plain, others have bright patterns, often with a light middorsal stripe. This species has large, spiny warts on the dorsal surface of the hind legs, especially the shanks; 1 or 2 warts in each large dark dorsal spot; well-developed cranial crests; and conspicuous, oblong parotoid glands. Each parotoid gland is separated from the transverse crest behind the eye by a short longitudinal ridge. Dark spots occur on the anterior part of the venter. Males are smaller than females, have horny tubercles on the first and second fingers, and have dark throats.

American Toads closely resemble Southern Toads and Fowler's Toads. The Southern Toad is best identified by the large knobs on its high cranial crests. Fowler's Toad has very low cranial crests, a more pointed snout, large dorsal spots enclosing three or more warts, and parotoid glands contiguous with the crests behind the eyes.

American Toads range throughout the Mountains and Piedmont of North Carolina and Virginia. They also occur in the Coastal Plain of north central and northeastern North Carolina and southeastern Virginia, and in the Mountains of South Carolina. These highly beneficial insectivores live in many habitats, from gardens to forests. In the Piedmont, they are abundant only during the cooler part of the year.

In our area, these are the earliest toads to breed—February or March in the south and March or April in the north. The mating call is a long, musical, whistlelike trill, often sustained for 20 to 30 seconds. A female lays about 6,000 eggs in two long strings on the bottoms of pools. Eggs hatch in less than a week, and metamorphosis occurs in about 2 months. The newly transformed toads are only 7 to 12 mm (0.3 to 0.5 in.) long.

When handled, a toad often discharges excess water stored in its cloaca. The parotoid glands and warts also secrete a white toxin that discourages many predators. The secretion can irritate a person's eyes and mouth, but toads can be handled without danger. They do not cause warts.

Fowler's Toad

Bufo fowleri (or *Anaxyrus fowleri*)
50 to 82 mm (2 to 3.25 in.) This
common, familiar toad has a brown,
olive, or gray dorsum with a middorsal light stripe. Each dark dorsal
spot contains three or more small
warts. The cranial crests are small
and contiguous with the parotoid
glands. The venter is whitish, usually
with a dark spot on the chest. Males
have black throats and are smaller

than females. Fowler's Toads sometimes hybridize with American Toads
or with Southern Toads, making
some specimens difficult to identify.

Fowler's Toad is abundant in
most habitats of the eastern United
States but is absent from the extreme southeastern corner of North
Carolina and most of the Coastal
Plain in South Carolina; it is uncommon or absent at higher elevations
in the Mountains.

Breeding occurs in March to May
in the south and April to July in the
north. From the edges of ponds,
lakes, and streams, a male emits a
loud, discordant "w-a-a-a-h" of about
1 to 4 seconds' duration. A female
lays about 7,000 eggs in two long
strings. The eggs hatch in about a
week, and the tadpoles transform
1 to 2 months later. The newly transformed toads are 8 to 11 mm (0.3 to
0.4 in.) long.

Oak Toad
Bufo quercicus (or *Anaxyrus quercicus*)

19 to 33 mm (0.75 to 1.25 in.) This dwarf among our toads is readily identified by a conspicuous light middorsal stripe and reddish orange tubercles on the undersurfaces of the hands and feet. The stripe is usually white but may be yellowish, buff, or reddish orange. The finely tuberculate back has 4 or 5 pairs of brown or black spots on a ground color of gray, brown, or nearly black, but these are obscure in darker toads. The dorsal tubercles are often reddish. The inflated vocal sac is ovate in shape.

Oak Toads inhabit grassy areas in pine savannas throughout most of the Coastal Plain south of the James River in Virginia. In South Carolina they occur in maritime forests and on some barrier islands.

This species breeds in the spring and summer, usually in response to heavy rains. The mating call is similar to the "peeps" of young chickens.

Females deposit about 700 eggs in beadlike chains of 2 to 8 eggs each in rain pools, ditches, or ponds. The transformed young are 7 to 8 mm (0.3 in.) long. Oak Toads often forage during the day for their small arthropod prey.

Once abundant in much of our area, Oak Toads have undergone serious population declines in recent years. Causes for these declines are not fully understood but may include habitat loss and alteration, disease, and/or predation by the Red Imported Fire Ant.

Southern Toad
Bufo terrestris (or *Anaxyrus terrestris*)
44 to 98 mm (1.75 to 4 in.) Prominent ridges and clublike knobs on the head distinguish this species from the American and Fowler's toads. Dorsal coloration is usually brown but may be red or blackish, sometimes with a light middorsal stripe. The dorsal dark spots usually contain 1 or 2 or more warts. The venter is grayish and the chest is spotted. A small, bony ridge separates the parotoid gland from the postocular ridge. Females are often paler than the slightly smaller, dark-throated males.

Southern Toads inhabit southeastern Virginia and all of the Coastal Plain in the Carolinas. They are abundant in a wide variety of habitats but seem to prefer areas with sandy, friable soils.

Most breeding occurs from late February to May, but some may take place earlier or later. The mating call is a long, whistlelike trill, similar to that of the American Toad but shorter in duration and an octave higher. This and our other two large *Bufo* species often utter a chirping "release call" if handled or mistakenly grasped by another male. Two long strings containing about 3,000 eggs are laid in shallow water. In a few days, the eggs hatch into tadpoles that metamorphose in 1 to 2 months; the toadlets are 7 to 10 mm (0.3 to 0.4 in.) long.

Northern Cricket Frog
Acris crepitans

16 to 35 mm (0.5 to 1.5 in.) Cricket Frogs have moist, warty skin and long legs with webs between the toes. There is a dark triangle between the eyes and a median stripe or a Y-figure on the back, which can be bright green, russet, yellow, or shades of brown or gray. The northern species differs from the southern by having a more rattling, variable, crepitant call; a more robust build; more webbing between the toes; usually a broader, less sharply defined dark stripe on the back of the thigh; and a pair of prominent, subanal, white tubercles.

This frog is predominantly a Piedmont species, but it also occurs in some portions of the northern and inner Coastal Plain and in a few localities along rivers in the lower Coastal Plain of South Carolina. It is very local in major valleys in the Mountains. It prefers open, grassy margins of ponds, ditches, and marshy areas.

Cricket Frogs breed in warm weather in shallow water. The call, a rapid "giick, giick, giick" resembling the clicking together of two small stones, is usually heard from about April to August. The eggs are laid singly or in small groups, attached to stems or scattered on the bottom. *Acris* tadpoles have a distinctive black tail tip. Transformation occurs in late summer when the froglets are about 14 mm (0.6 in.) long. Small insects and spiders comprise the diet.

Southern Cricket Frog
Acris gryllus

16 to 32 mm (0.5 to 1.5 in.) This species is very similar to the Northern Cricket Frog but is slightly smaller and more slender. The snout is more pointed, the legs are longer with less webbing between the toes, and there is a sharply defined, narrow, dark stripe on the back of the thigh. The subanal tubercles are usually indistinct. The two species are most reliably separated by their calls.

The Southern Cricket Frog

replaces the northern species in most of the Coastal Plain, but the two are sympatric in some areas. A better jumper than the Northern Cricket Frog, this species is abundant along the grassy margins of quasi-permanent ponds, streams, and ditches. Water lily meadows are especially preferred habitats.

Though this species is active throughout the year, it breeds mostly in late spring and summer. Its clicking call, a slightly less variable and more metallic "gick, gick, gick" than that of the Northern Cricket Frog, may be heard by day or night from about February to October. The eggs and tadpoles are similar to those of the Northern Cricket Frog. About 150 eggs are laid at a time, and more than one complement may be produced each year. The Southern Cricket Frog may be declining in the northern part of its range in our area. The ecological relationships between it and the Northern Cricket Frog deserve more study. The two are not known to hybridize.

Pine Barrens Treefrog
Hyla andersonii

29 to 51 mm (1 to 2 in.) This beautiful frog is easily identified by the distinctive purple stripe along each side of the body. The dorsum is uniformly green, the venter is white, and the concealed surfaces of the legs are bright orange with numerous small yellow spots.

This uncommon species is known from the Sandhills and south central Coastal Plain of North Carolina and the upper Coastal Plain in South Carolina. It inhabits bayheads, pocosins, seeps, shrub bogs, pitcher plant wetlands, and similar habitats in and adjacent to sandhills.

Calling has been reported from March to September, but most breeding takes place from April to July. The call is a high, nasal "quonk," quickly repeated 10 to 20 times at irregular intervals. Males may call from the ground or from shrubs or low trees near bogs, ditches, or slow-moving streams and are difficult to locate. In shallow water, each female lays about 500 eggs, usually singly or in small clusters. They develop rapidly and hatch in 3 or 4 days. The newly transformed frogs are about 15 mm (0.6 in.) long.

The Pine Barrens Treefrog is listed as a species In Need of Management in South Carolina.

Bird-voiced Treefrog *Hyla avivoca*
25 to 51 mm (1 to 2 in.) This frog is
similar in general proportions and
pattern to the two species of Gray
Treefrogs. It differs, however, by
averaging slightly smaller and by
having pale greenish or yellowish
white pigment, rather than bright
orange mottled with black, on the
concealed surfaces of the hind legs
and groin.

Bird-voiced Treefrogs inhabit
heavily wooded swamps bordering
blackwater rivers and streams. In
our area they are known only from
the swamps of the Savannah River
and its tributaries in Aiken, Allendale, Barnwell, Hampton, and Jasper
counties, South Carolina.

In late spring and summer,
males call from heights of 1 to 2 m
(3 to 6.5 ft.) on bushes and trees in
or near water. The call is distinctive—a loud, rapidly repeated, birdlike whistle: "whit, whit, whit." It is
reminiscent of an osprey's call or of
someone whistling for a dog. In shallow water, a female deposits about
650 eggs in packets of 6 to 15. At high
temperatures, they hatch in about 40
hours, and metamorphosis occurs
in 31 to 33 days. Newly transformed
frogs are about 13 mm (0.5 in.) long.

Cope's Gray Treefrog
Hyla chrysoscelis

32 to 62 mm (1.25 to 2.5 in.) This common, familiar treefrog is usually some shade of gray, but an individual may vary from whitish to pale brown to green. The relatively rough, moist skin is usually patterned with dark, irregular markings that offer superb camouflage against lichen-covered tree bark. There is a conspicuous whitish patch under the eye and bright orange or yellow on the concealed surfaces of the hind legs. The belly is whitish with granular skin, and males have dark throats. This and the Gray Treefrog are sibling species, identical in appearance but differing genetically. The diploid Cope's Gray Treefrog has only 24 chromosomes; the tetraploid Gray Treefrog has 48. This difference is readily revealed by microscopic examination of cells of the inner eyelid (nictitating membrane). A cell of *H. chrysoscelis* has 1 or 2 nucleoli; that of *H. versicolor* has 3 or 4. Aside from such laboratory techniques, the two may be distinguished in the field by their calls and, in many cases, by their geographic location. The similar Bird-voiced Treefrog may be distinguished by the pale green or yellowish white (rather than bright yellow or orange) on its hind legs.

Cope's Gray Treefrogs range throughout most of our region. They forage for insects amid trees and shrubs and on the ground. Cryptic and well camouflaged, they are not often encountered outside the breeding season (April to August); but they may be found on roads on

rainy nights, and sometimes individuals visit lights at night to feed on insects.

The call is a short, vibrant, flute-like trill. Shorter, harsher, and more forceful than that of the Gray Treefrog, it contains an average of 45 trills per second. Scattered packets of 10 to 40 eggs are laid on the surface of ditches, puddles, shallow ponds, and other usually temporary wetlands. Hatching occurs in 4 to 5 days, and metamorphosis takes 45 to 64 days. The recently transformed young are about 13 to 20 mm (0.5 to 0.75 in.) long and are often more greenish than the adults.

Green Treefrog *Hyla cinerea*
32 to 64 mm (1.25 to 2.5 in.) This large, slender treefrog has long legs and smooth skin. The venter is plain white; the dorsum is usually bright green or yellowish green but occasionally dark greenish brown or olive. Most specimens have tiny golden spots on the back and a prominent light stripe on each side of the body. The stripe may be reduced or absent, especially in northern populations.

The Green Treefrog occurs throughout the Coastal Plain in our region and has expanded its range far into the Piedmont in recent years. More fish-tolerant than our other treefrogs, it often breeds in permanent water and prefers the floating and emergent vegetation along the swampy edges of ponds, lakes, marshes, and streams. During the day, these well-camouflaged frogs rest motionless, often on cattails or other wetland plants; at night, they are sometimes attracted to insects near lights.

Most spawning occurs in May or June, but the mating call may be heard from April to September. The call is a bell-like, nasal "queenk," repeated once a second. Each female lays about 400 eggs amid floating vegetation. The larval stage lasts about 2 months, and the newly transformed frogs are 12 to 17 mm (0.5 to 0.7 in.) long.

Pine Woods Treefrog
Hyla femoralis

25 to 38 mm (1 to 1.5 in.) This slender treefrog is usually brown or reddish brown with dark markings on its back, but it may vary from pale gray to dark brown to greenish. Individuals may change color in response to temperature, light, and moisture. Small yellow or orange spots against a dark background on the rear of the thigh provide a diagnostic feature. A dark, ragged stripe runs from the nostril through the eye and down the side of the head and body. The Bird-voiced and the two Gray Treefrogs are similar, but these species have a prominent white spot just below the eye.

This well-named species lives in pine flatwoods and savannas, usually near ephemeral wetlands. It is occasionally found in hardwood forests and swamps. Pine Woods Treefrogs occur throughout most of the Coastal Plain and portions of the adjacent Piedmont.

The Pine Woods Treefrog breeds in temporary wetlands from March to October, most often after heavy rains. The mating call is a low-pitched, guttural trill that sounds somewhat like "getta, getta, getta" and has often been compared to Morse code. It is often heard in pine savannas on overcast days. Females deposit films of 100 to 125 eggs on the water's surface or just below it on stems or other objects. The tadpole stage lasts 50 to 75 days; transformed young are about 13 mm (0.5 in.) long.

Barking Treefrog *Hyla gratiosa*
51 to 69 mm (2 to 2.75 in.) Our
largest treefrog, this stout, attractive
frog usually has prominent circular
dorsal spots on a background of
pale greenish gray to bright green or
dark brown. In the extremely light
or very dark phases, the dorsal spots
may be obscure. Individuals may
undergo color changes in response
to temperature, light, and humidity.
The belly is white or yellowish white,
and the skin has a granular texture.
A thin white stripe usually borders
the upper lip and extends onto the
anterior sides of the body, and the
outer surfaces of the forelimbs and
hind feet are similarly bordered with
white. Males may be slightly larger
than females, a phenomenon unusual among our native frogs.

Barking Treefrogs are common
locally in sandy areas near shallow
ponds in pine savannas and in low,
wet woods and swamps. They occur
throughout much of the Coastal
Plain and in portions of the adjacent
Piedmont of the Carolinas, and there
are a few imperiled populations in
southeastern Virginia.

This species is most active in the
late spring and summer. Breeding
occurs in shallow ponds after heavy
rains. As the males move into the
ponds from bushes or trees, they
utter series of loud, distinctive, dog-
like barks, but after they enter the
water, their call changes to a hollow-
sounding "doonk" repeated at short
intervals. Males call while floating
on the water's surface. Females
deposit eggs singly on the bottom of
the pond; the total egg complement
is about 2,000. The tadpole stage
lasts 40 to 70 days; transformed
young are 14 to 20 mm (0.6 to 0.8 in.)
long.

The Barking Treefrog is listed as
a Threatened species in Virginia.

Squirrel Treefrog *Hyla squirella*
22 to 41 mm (1 to 1.5 in.) This small
treefrog can rapidly change its color
and markings. The dorsum may be
dull brown with many or few round
dark spots, or it may be light or
bright green and unspotted. There
is at least a partial bar between the
eyes and a whitish line on the upper
lip, shoulder, and side of body. The
belly is whitish with a suffusion of
yellow on the axilla, groin, thigh,
and shank. This species differs from
other *Hyla* in that it lacks a light spot
under the eye and yellow or orange
spots on the thigh.

This common frog prefers open
woods but often occurs around
buildings in cities and towns. It
ranges from southeastern Virginia
throughout the Coastal Plain of the
Carolinas and sporadically in the
adjacent Piedmont, where it may
have undergone some range expansion in recent years.

Most breeding occurs from April
to August and is usually associated
with summer storms. The call is a
flat, nasal, ducklike "waaaak," 0.25
seconds long and repeated every 0.5
seconds. A female lays about 1,000
eggs on the bottom of open, usually
ephemeral ponds or pools. Tadpoles
transform after 45 days, and the
new frogs average 12 mm (0.5 in.)
from snout to vent. Small insects are
the chief food, and these frogs are
commonly observed feeding around
lights at night.

Gray Treefrog *Hyla versicolor*
32 to 62 mm (1.25 to 2.5 in.) This
treefrog is virtually identical in
appearance to its sibling species,
Cope's Gray Treefrog. It has, however, twice as many chromosomes;
a slower, lower-pitched call; and a
more northerly range.

Gray Treefrogs occur over much
of the Mountains and Piedmont of
Virginia. In North Carolina they have

been documented from Warren and
Caswell counties along the Virginia
border and are suspected to occur
in some other localities in the upper
Piedmont and Mountains. They have
not been reported from South Carolina. Their habits are similar to those
of Cope's Gray Treefrog.

The call is a vibrant, flutelike
trill. Longer, more melodic, and less
forceful than that of Cope's Gray
Treefrog, it contains an average of
25 trills per second. Gray treefrogs
breed in ditches, puddles, shallow
ponds, and other temporary wetlands. Small masses of 10 to 40 eggs
are laid on the water's surface. These
hatch in about 4 to 5 days, and metamorphosis occurs in 45 to 65 days.
Metamorphs are about 13 to 20 mm
(0.5 to 0.75 in.) long and are often
more greenish than the adults.

Mountain Chorus Frog
Pseudacris brachyphona

25 to 38 mm (1 to 1.5 in.) The brown or gray dorsum of this small frog has a dark mark between the eyes and a pair of irregular lateral markings, often similar to "reversed parentheses." These are sometimes connected near the midback to form a crude H or X. The granular ventral skin is whitish, and males have dark throats. This species differs from other chorus frogs by having a stockier body, a wide head, large digital pads, and no middorsal stripe or spots. It may resemble the Spring Peeper but has smaller toe pads and a white line on the upper lip.

Mountain Chorus Frogs inhabit wooded areas in western Virginia and in Cherokee County, North Carolina.

Breeding occurs from February to April. The mating call is a harsh, raspy "wreenk" or "reek" and may be heard day or night. The 400 eggs in the average complement are laid in groups of 10 to 50 and are attached to vegetation in shallow, quiet ponds, ditches, and small pools along the edges of streams. They hatch in 4 or 5 days into tadpoles about 5 mm (0.2 in.) in total length. Tadpoles reach a total length of 30 mm (1.2 in.) and transform in 50 to 56 days. The froglets average 8 mm (0.3 in.) long.

The Mountain Chorus Frog is listed as a species of Special Concern in North Carolina.

Brimley's Chorus Frog
Pseudacris brimleyi

25 to 32 mm (1 to 1.25 in.) This small, tan frog has three brown stripes on its back and a distinct dark brown or black stripe on each side extending from the nostril through the eye to the groin. On some individuals the dorsal stripes may be obscure. The undersurface is yellowish, and the chest is usually spotted with brown.

This species is readily distinguished from the Coastal Plain form of the Upland Chorus Frog by the horizontal brown stripe on the rear of its shank.

Brimley's Chorus Frogs inhabit low areas in hardwood forests and swamps near rivers and streams. They occur from southeastern Virginia southward throughout much of the Coastal Plain of the Carolinas.

This secretive and seldom-seen species breeds in winter and early spring. The mating call is a short, guttural trill, somewhat similar to that of the Squirrel Treefrog. Females deposit small, loose clusters of eggs on stems or other objects in ditches, shallow ponds, or other temporary wetlands. The tadpole stage probably lasts 40 to 60 days; transformed young are 9 to 11 mm (0.4 in.) long.

Spring Peeper *Pseudacris crucifer*
19 to 35 mm (0.75 to 1.5 in.) This
familiar harbinger of spring has a
tan, brown, orange, or gray dorsum
with a prominent dark X-shaped
marking near the middle of the back
and a dark, barlike marking between
the eyes. The underparts are buff,
cream colored, or yellowish, and
sometimes spotted. Males have dark
throats and are usually smaller and
darker than females.

Spring Peepers occur throughout
the area, with the exception of most
barrier islands. They inhabit woodlands and live under forest litter or
amid brushy undergrowth. Adults
are secretive and much more often
heard than seen.

Breeding occurs in woodland
ponds, swamps, and ditches from
October to March in the southern
coastal regions and from February to
June in the northern and mountainous areas. The call—a clear, high-
pitched (birdlike) "peep" ending
with an upward slur—is repeated
about once a second. In some portions of our region, Spring Peepers
may be heard during all months of
the year. The eggs (about 900 per
complement) are attached singly to
submerged objects. They hatch in a
few days, and metamorphosis occurs
3 to 4 months later. Spring Peepers
feed mostly on small insects and
other arthropods.

Upland Chorus Frog
Pseudacris feriarum

19 to 35 mm (0.75 to 1.5 in.) This frog's dorsum is brown or gray with a variable pattern. A distinct dark stripe extends from snout through eye to groin, and the three dorsal stripes may be reduced to rows of spots. A dark triangular marking usually lies between the eyes, and a light stripe lines the upper lip. The venter is granular and cream colored, often with dark stippling on the chest. The toes have small pads and little or no webbing. This species is sometimes treated as a subspecies of the Western Chorus Frog, *P. triseriata.*

Upland Chorus Frogs occur at low elevations in the Mountains, widely throughout the Piedmont and upper Coastal Plain, and in scattered populations in the lower Coastal Plain of North Carolina.

Breeding takes place in woodland pools, ditches, and other temporary wetlands from December to March in the south and February to May in the north. The mating call is a regularly repeated "crrreek," imitated by thumbnailing the teeth of a comb. A female lays about 1,000 eggs, in clusters of about 60 each, attached to vegetation. The tadpole state lasts 2 to 3 months; average size at metamorphosis is 9.5 mm (0.4 in.). These secretive forest-floor dwellers eat small arthropods. Though abundant in some areas, they are seldom encountered outside the breeding season.

New Jersey Chorus Frog
Pseudacris kalmi

19 to 39 mm (0.75 to 1.5 in.) Three usually unbroken, broad, black to dark gray stripes down the brown or gray back characterize this small frog. There is often a narrow, dark line running from the snout through the eye to the groin and a dark, triangle-shaped mark between the eyes. A white to cream line borders the upper lip. Juveniles are similar to adults, but their stripes are less pronounced. This frog is sometimes treated as a subspecies of the Western Chorus Frog, *P. triseriata*. Rely on geographic location to help identify it; the Spring Peeper is the only other chorus frog within its range.

This species occurs on the Delmarva Peninsula northward to Staten Island, New York. Populations in the Virginia portion of the peninsula occur in Accomack and upper Northampton counties. They are seldom seen because of their tendency to hide in grasses in shallow water with only their snouts protruding above the water. These chorus frogs inhabit hardwood and mixed forests with a shrub understory and shallow-water breeding sites nearby.

These frogs breed over a short period from January to March, depending on the temperature. Males will call for about 2 to 3 weeks, but females will be present at breeding sites for only 3 to 4 nights. Their call, corresponding to the sound made when a finger is run over a stiff comb, is similar to that of the Upland Chorus Frog. Females lay about 30 to 100 eggs in several small, clear masses attached to stems in shallow water. The larval period lasts about 3 to 4 weeks. Metamorphs are about 12 to 14 mm (0.5 in.) in body length. Little is known about reproduction and population dynamics in this species.

Southern Chorus Frog
Pseudacris nigrita

19 to 32 mm (0.75 to 1.25 in.) This small, grayish or brownish frog has a prominent black stripe on each side from the snout to the groin and three rows of irregular black spots on the back. The belly is usually white. A distinct silvery or yellowish white stripe usually extends along the upper lip. Upland Chorus Frogs are similar but have yellowish bellies spotted with brown, a tan or brown back, and often a dark triangular spot between the eyes (which may be absent in Coastal Plain frogs).

This pine savanna species breeds in ditches, Carolina bays, shallow ponds, and other temporary wetlands. It occurs throughout much of the Coastal Plain in the Carolinas, and disjunct populations were recently discovered in several counties in southeastern Virginia. Adults are terrestrial, secretive, and seldom encountered when not breeding.

Southern Chorus Frogs breed from late fall to early spring. The mating call is a mechanical, ascending trill, similar to a ratchet wrench or the sound produced by scraping the teeth of a comb. Females deposit small, irregular egg clusters on stems, leaves, or other objects in shallow water. The tadpole stage lasts about 50 days; recently transformed young are about 9 to 15 mm (0.3 to 0.6 in.) long. Chorus frogs are short lived, and populations are vulnerable to prolonged drought, habitat alteration, and other large-scale disturbances. This species has undergone apparent declines in recent years.

Little Grass Frog
Pseudacris ocularis

13 to 19 mm (0.5 to 0.75 in.) This diminutive frog, the smallest in North America, varies from gray to brown or brick red. A dark stripe extends from the nostril through the eye onto the sides, and a dark mid-dorsal stripe is sometimes present. The belly is yellowish white. In late spring, Little Grass Frogs can be confused with the young of Brimley's and Southern Chorus Frogs, but inflated vocal sacs in males or eggs showing through the belly in females will help distinguish this species.

Little Grass Frogs prefer grassy areas near bogs or ponds in pine savannas, and pools or streams in hardwood forests and swamps. They occur in most of the Coastal Plain.

More often heard than seen, this common species usually breeds in association with late winter, spring, or summer rains. Its insectlike call — a "pt-seee" so high-pitched that some persons have difficulty hearing it — may be heard in warm weather at almost any time of year. Females deposit about 100 eggs singly on the bottom of shallow ponds or other temporary wetlands. Transformation occurs after a tadpole stage of 45 to 70 days; recently transformed young are 7 to 9 mm (0.3 in.) long.

Ornate Chorus Frog
Pseudacris ornata

25 to 38 mm (1 to 1.5 in.) This colorful frog may be tan, green, or brick red. The concealed surfaces of the thigh and groin are usually bright yellow. A dark, triangular spot is often present between the eyes, and there are prominent black spots along the sides, on the lower back, and in the groin. A bold black stripe extends from the nostril through the eye to the shoulder on each side.

Ornate Chorus Frogs primarily inhabit Longleaf Pine savannas; breeding males call from Carolina bays, ditches, and other ephemeral wetlands. They occur locally in the Coastal Plain from Beaufort County, North Carolina, southward.

This increasingly uncommon species breeds mostly in winter. Its call — a series of loud, metallic, birdlike peeps resembling the trill of the Spring Peeper but lacking its ascending quality — may be heard between October and March. Females deposit small, irregular clusters of 10 to 100 eggs on stems or other objects in shallow water. The recently transformed young are 14 to 16 mm (0.6 in.) long.

Like many upland ephemeral pond breeders associated with the Longleaf Pine ecosystem, this frog has undergone serious population declines in recent years.

Eastern Narrow-mouthed Toad
Gastrophryne carolinensis
22 to 38 mm (1 to 1.5 in.) These small, stocky toads have smooth, moist skin; a pointed snout; and a unique fold of skin across the back of the head. The dorsum is gray, brown, or reddish with a broad, dark central marking. The venter is heavily speckled or mottled, and males have dark throats. The absence of parotoid glands, warts, visible tympani, and webs between the toes distinguishes these toads from other species.

Narrow-mouthed Toads occur throughout the Coastal Plain and lower Piedmont but are absent from most of the Mountains and upper Piedmont. Adults are terrestrial and secretive, spending much time in burrows or under logs, rocks, or other sheltering objects. They are probably the most specialized feeders among our frogs, taking mostly small ants and often positioning themselves near anthills for feed-

ing. Termites and some other small insects are also eaten.

During warm rains in April to October, large numbers aggregate in shallow puddles, ponds, ditches, flooded fields, and other often temporary wetlands. The mating call is a prolonged, lamblike, nasal "baaaa." Packets of eggs are laid on the surface of the water; each complement contains about 850 eggs. In 20 to 70 days, metamorphosis occurs; the new frogs measure 8.5 to 12 mm (0.3 to 0.5 in.) long.

Carolina Gopher Frog
Rana capito (or Lithobates capito)
72 to 94 mm (2.75 to 3.5 in.) This large, toadlike frog has a large head, wide mouth, and cobblestone rows of prominent warts on its back. The belly is heavily mottled with dark spots or flecks, and the concealed surfaces of the thigh and groin are suffused with yellow. The back is dark gray with relatively prominent dorsolateral folds.

Gopher Frogs inhabit dry, Turkey Oak–pine associations and other sandy areas in pine savannas. They are uncommon to rare, highly terrestrial, and enter water only to breed. When not active on the sur-

face, these secretive, nocturnal frogs occupy stump holes or the burrows of crayfish and other animals. The range of this species extends north in the Coastal Plain to Carteret and Jones counties (historically to western Beaufort County), North Carolina.

Gopher Frogs breed in upland ephemeral ponds in pine savannas after heavy rains. Most breeding takes place from mid-February to mid-April, but some breeding may occur in fall. The mating call, a deep, humanlike snore, is sometimes given underwater. The egg mass is a large globular cluster attached to stems. The tadpole stage lasts 85 to 100 days; transformed young are 27 to 38 mm (1.1 to 1.5 in.) long. When handled, a Gopher Frog usually arches its body and places its forelimbs over its face, apparently to protect its head while exposing the toxic glandular skin on its back.

The Carolina Gopher Frog is listed as an Endangered species in South Carolina and as a Threatened species in North Carolina.

American Bullfrog
Rana catesbeiana (or Lithobates catesbeianus)

85 to 200 mm (3.5 to 8 in.) This familiar frog is North America's largest. The dorsum is olive, green, or brown with large, obscure blotches on the adults and many small black dots on the young. The venter is buffy white with dark reticulations or mottling. Bullfrogs lack a dorsolateral fold but have a thin fold alongside the tympanum. Males are smaller than females and have a yellow throat and larger tympanum, thumb, and forearm. River Frogs have rougher skin and light spots on the upper lip. Pig Frogs have a more pointed snout and more webbing between the toes.

Bullfrogs are common in the Carolinas and Virginia, occurring virtually throughout the region. Highly adaptable, they may live in almost any body of freshwater; but adults prefer large ponds, lakes, and streams, and most successful breeding is in permanent water.

The vibrant, deep bass "jug-o-rum" can be heard over great distances during late spring and early summer. About 12,000 eggs are laid by each female. They hatch in about 5 days, and metamorphosis usually requires a year—sometimes longer. Some tadpoles may grow as large as 125 to 150 mm (5 to 6 in.) before transforming. Bullfrogs take virtually any prey items they can overpower and swallow, including insects, crayfish, and small vertebrates.

Green Frog
Rana clamitans (or *Lithobates clamitans)*

54 to 86 mm (2 to 3.5 in.) The Green Frog's dorsolateral folds extend only to the middle of the back. The dorsum may be green but is more often brown or bronze, usually with obscure dark spots and at least some bright green on the face and upper lips. The venter is white, but dark markings usually occur on the chin, breast, and hind legs, especially on young frogs. The male has a large tympanum, yellow throat, and stout

foreleg and thumb. Most specimens from the Coastal Plain of South Carolina and southeastern North Carolina average slightly smaller, may have more dark vermiculations on their bellies, and have plain brown or bronze backs, often with little or no green on the upper lips.

Green Frogs are common along streams, ponds, and lakes throughout the Carolinas and Virginia. A high-pitched "squeenk" is often given as a startled frog jumps to safety.

The call—a low, explosive, twangy "c'tung," often likened to a loose banjo string being plucked—may be heard from March to September, but most breeding is from April to June. About 3,000 eggs are laid in a raftlike surface film. Most tadpoles transform in a few months, but some overwinter. Sexual maturity occurs near the end of the first full summer after metamorphosis. Mainly arthropods, snails, worms, and other invertebrates are eaten.

Pig Frog
Rana grylio (or Lithobates grylio)

80 to 162 mm (3 to 6.25 in.) Pig Frogs resemble Bullfrogs but have a more narrow and pointed snout, conspicuous light and dark horizontal bands on the rear of the thigh, and more fully webbed hind feet. Young Pig Frogs are similar to adult Carpenter Frogs. To distinguish them, note that the Pig Frog's longest toe is webbed nearly to its tip, but in Bullfrogs and Carpenter Frogs the webbing extends only to the midpoint of the last joint.

This species prefers large, open, relatively shallow ponds or lakes with lily pads and much emergent vegetation. It is abundant in abandoned rice fields and rice field reservoirs. In our region, Pig Frogs occur only in the South Carolina Coastal Plain north to Georgetown and Clarendon counties.

The common name of this frog stems from its distinctive mating call, an explosive, piglike grunt; breeding choruses occur from early April to early August. Each female deposits about 10,000 eggs in a thin layer on the surface of the water. The tadpole stage lasts about 1 year; transformed young are 32 to 49 mm (1.25 to 2 in.) long.

River Frog
Rana heckscheri (or Lithobates heckscheri)

80 to 135 mm (3 to 5.25 in.) This large, dark brown frog has conspicuous whitish spots on the margins of its jaws, a dark belly with pale mottling, and often a pale band around the groin. It may be confused with dark Bullfrogs; but these have paler bellies and usually at least some green on the face, and their jaws lack white spots. Some Green Frogs may look similar, but they have dorsolateral folds.

River Frogs inhabit oxbow lakes, ponds, borrow pits, river swamps, and other permanent waters associated with blackwater rivers. They occur in the South Carolina Coastal

Plain north to Sumter and southern Marion counties. They were known historically from the Lumber and Cape Fear rivers in North Carolina, where they have not been reported in over 30 years and are believed extirpated.

This species breeds in late spring and summer. The mating call is a deep, trainlike snore, but territorial males may also voice a loud, explosive grunt. Females deposit eggs as a surface film. The very large, conspicuous tadpoles reach lengths of 150 mm (6 in.), have tail fins conspicuously edged with black, and aggregate in large, fishlike schools—a behavior unique among North American tadpoles. The tadpole stage lasts 1 or 2 years. Recently transformed young are 30 to 52 mm (1.2 to 2.1 in.) long and resemble adults but are more olive colored and usually have red irises. River frogs produce skin toxins unpalatable to many predators. Less wary than our other ranids, they often go limp when handled.

Although of uncertain occurrence in North Carolina, the River Frog is listed as a species of Special Concern there.

Pickerel Frog
*Rana palustris (*or *Lithobates palustris)*

44 to 87 mm (1.75 to 3.5 in.) This moderately large frog has distinct dorsolateral folds and two rows of squarish spots on the back. In Coastal Plain individuals, the spots often merge to form short, oblong bars or broad, elongate stripes. The belly is white in Mountain and Piedmont frogs and mottled with brown in Coastal Plain ones. The concealed surfaces of the thigh and groin are bright yellow to orange. Southern Leopard Frogs have oval spots on the back and no yellow or orange on the thigh and groin.

This species utilizes a variety of aquatic habitats in wooded areas, including bogs and grassy places near shaded streams. The wary, agile adults may travel far from water. They occur throughout most of the Carolinas and Virginia, but populations are widely disjunct.

Most Pickerel Frogs breed during the late winter and early spring, usually with the advent of heavy rains. The mating call is a low-pitched snore, sometimes given underwater. Females deposit globular clusters of about 2,500 eggs on stems in ponds or pools. The tadpole stage lasts 70 to 80 days; transformed young are 19 to 27 mm (0.75 to 1.1 in.) long. Pickerel Frogs produce skin secretions that are toxic or distasteful to many predators.

Southern Leopard Frog
*Rana sphenocephala (*or *Lithobates sphenocephalus)*
50 to 90 mm (2 to 3.5 in.) The dorsum of this long-legged, active frog is green or brown or both, with large, rounded dark spots; in contrast, the venter is white. Distinct dorsolateral folds extend the full length of the body. There is a white spot in the center of the tympanum. Males are smaller than females and have paired vocal sacs and enlarged forearms and thumbs. The Pickerel Frog is similar but has more angular dorsal spots, lacks a white spot on the tympanum, and has bright yellow or orange on the concealed surfaces of the hind legs.

Southern Leopard Frogs are most

abundant in the Coastal Plain but also range well into the Piedmont and penetrate into the Mountains of South Carolina. They inhabit ponds, ditches, and swamps, as well as the margins of lakes and streams. Powerful leapers, they forage extensively on land, primarily for insects, and often travel far from water.

This species usually breeds in winter or early spring but may do so at almost any time of year. The call is a series of 3 to 5 guttural croaks followed by 2 or 3 clucks. It suggests the sounds made by rubbing an inflated balloon. Calling males are wary and difficult to approach. Each female attaches a firm cluster (about 90 mm [3.5 in.] wide and 40 mm [1.5 in.] thick) of several hundred eggs to vegetation just below the water's surface. Breeding Leopard Frogs often congregate and lay numerous clusters of eggs in a small area, often in very shallow water. These hatch in a week or two. About 3 months later, the average tadpole is 65 to 70 mm (2.5 to 2.75 in.) long, its tail bears prominent dark spots, and metamorphosis is imminent. The newly formed frogs are about 20 mm (0.8 in.) long.

Wood Frog
Rana sylvatica (or Lithobates sylvaticus)

35 to 83 mm (1.5 to 3.25 in.) This medium-sized frog is easily identified by its mask, a dark patch extending back from the eye. The dorsum varies from light tan to reddish to dark brown. The females are more brightly colored and larger than the males. The venter is white with a dark spot on each side of the chest. Dorsolateral folds are prominent.

Wood Frogs live in or near moist woods, often far from open water. In the Carolinas, they occur in the Mountains and upper Piedmont and in disjunct relict populations in Hyde and Tyrrell counties in the outer Coastal Plain of North Carolina; in Virginia, they are scattered throughout the Mountains and Piedmont as well as the northern part of the upper Coastal Plain.

Wood Frogs are highly tolerant of freezing temperatures. Breeding is usually limited to a few days in January, February, or March when large numbers congregate after winter rains. The mating call, a rasping "craw-aw-auk," has little carrying power. The globular egg masses are often closely aggregated and attached to plants below the surface of a shallow pond or pool. After 40 to 50 days, the tadpoles transform. These frogs feed chiefly on beetles, flies, and other insects. They hibernate under leaves or logs in wooded ravines.

Carpenter Frog
Rana virgatipes (or Lithobates virgatipes)

41 to 67 mm (1.5 to 2.5 in.) This small frog has 4 yellowish brown stripes, 2 on the back and 1 on each side. The rear of the thigh has alternating light and dark stripes, and the belly is usually mottled with black. Dorsolateral folds are lacking. Young Pig Frogs are similar, but their longest toe is webbed nearly

to its tip; in Carpenter Frogs, two joints of the longest toe are free of webbing.

More often heard than seen, Carpenter Frogs are wary and blend in well against sphagnum mats and other vegetation in the coffee-colored waters of the pine savanna bogs or ponds in which they live. Our most highly aquatic frog species, they occur throughout most of the Coastal Plain.

Carpenter Frogs breed in spring and summer. The mating call, an explosive, two-noted "clack-it," is repeated 3 to 6 times in rapid succession and resembles the sound of a distant carpenter hammering. The egg mass is a flattened or globular cluster containing 200 to 600 eggs. The tadpole stage lasts about 1 year; transformed young are 23 to 31 mm (0.9 to 1.2 in.) long. Their diet consists of many types of invertebrates.

Class Reptilia

Reptiles evolved from primitive amphibians about 300 million years ago. The oldest fossils are from the Pennsylvanian period of the Paleozoic era. Reptiles were the first vertebrates to become completely free of the aquatic environment. By the Mesozoic era, they had become the dominant vertebrates and radiated not only in terrestrial environments but also in marine, freshwater, and aerial habitats. In the Mesozoic, these highly successful animals gave rise to the birds and mammals, but by the close of that era, 12 of the 16 orders had become extinct. Only 4 orders are living today: Testudines (turtles); Crocodilia (crocodiles, alligators, and their kin); Rhynchocephalia (tuataras — two relict species native to the New Zealand area); and Squamata (lizards and snakes).

Reptiles have a dry, glandless skin covered with horny scales. Snakes and a few lizards are legless, but basically reptiles have 2 pairs of limbs, each usually with 5 clawed digits. Fertilization in all reptiles is internal. Some snakes and lizards give birth to live young, but most reptiles lay large eggs provided with an enormous yolk and a protective shell. Unlike the eggs of fish and amphibians, the reptile egg is specialized for development on land. In addition to the *yolk sac*, a primitive membrane that permits a vertebrate embryo to use nutrients stored in the egg, reptiles and their descendants have 3 additional membranes: the *amnion* surrounds the embryo and secretes a fluid that suspends and protects the embryo; the *allantois*, a large saclike extension of the digestive tube, is vascular and stores metabolic wastes and is also a respiratory surface; and the *serosa* encloses the developing young and the other membranes. The shell is leathery or calcium impregnated. Its porosity allows exchange of respiratory gases and the uptake of water. It protects the embryo from physical pressures and desiccation. Unlike fish and many amphibians, reptiles lack gills, have no free-living larval stage, and breathe solely by means of lungs. In many species the females store sperm for long periods; some have produced viable young 5 years after copulation.

About 6,000 species of reptiles are living today. They are most numerous in the tropics and subtropics, but a few live near the Arctic Circle and at high elevations—up to 4,900 m (16,070 ft.) in the Himalayas and the Andes Mountains. Even though they are not warm blooded, many reptiles maintain a surprisingly high body temperature within narrow limits. This is accomplished mainly by behavioral means such as basking, orientation, postural changes, and habitat selection; however, some also use limited physiological means: panting, vasomotor responses, and metabolic heat production.

Order Crocodilia

CROCODILES AND ALLIGATORS

About 200 million years ago the Crocodilia evolved from the Thecodontia (Triassic reptiles with teeth firmly set in sockets). They flourished in the Jurassic and Cretaceous but now include only 21 living species. Modern crocodilians are relatively similar in general body shape and in habits. They are medium to very large, the adults ranging in total length from 1 m [3.2 ft.] to almost 7 m [23 ft.]. All have long, powerful jaws equipped with large teeth set in deep sockets; a long, muscular tail; and 2 pairs of limbs. Each front foot has the usual 5 toes, but each hind foot has only 4 toes. Very large, horny scales cover the body, and many, especially those on the back, contain plates of dermal bone. Crocodilians are well equipped for semiaquatic living; the nostrils and ear openings are valvelike and close when submerged, and the heart rate may drop from about 28 to 5 beats per minute. A diaphragmlike septum separates the lungs from the abdominal viscera, and a bony palate separates the nasal passage from the mouth. The heart in most modern species has a complete septum between the ventricles, as occurs in birds and mammals. Indeed, modern crocodilians are more closely related to birds than to other living reptiles. Many systematists advocate grouping them in a separate class. For the purposes of this book, we have retained the traditional classification of regarding crocodilians as an order within the Reptilia.

These predators eat a wide variety of organisms, ranging from small invertebrates to large mammals. Some crocodilians are territorial, and many species produce a deep, rumbling bellow. Males have a single median copulatory organ. Most species lay 14 to 100 eggs per clutch in mound-shaped nests constructed of soil and vegetation; however, a few dig holes for their eggs. Unlike most other reptiles, crocodilians often actively defend the eggs and the very young from predators. In some species, the female assists the hatchlings by opening the nest and gently carrying them in her mouth to water.

Because their skins are commercially valuable and because a few of the larger species may be dangerous to humans, crocodilians have been eagerly hunted. This persecution has been exacerbated by continual destruction of their habitat. Some species are perilously close to extinction, and for some of them, recent protective measures may be too late. In contrast, the American Alligator has shown great resiliency under protection during the last few decades. American Alligators have become so abundant that they are a nuisance in some parts of more southern states.

Crocodilians are chiefly tropical. Only the American and Chinese alligators occur in the north temperate zone. The former is the only crocodilian native to our area. The Spectacled Caiman has been extensively imported from Central and South America for the pet trade. After the novelty wears off, many of these animals are released. Although some may survive for several years, there is no evidence that any reproduce in our area.

American Alligator
Alligator mississippiensis

1.8 to 5.8 m (6 to 19 ft.) This large, unmistakable reptile has a broad snout, a short neck, a heavy body, and a laterally compressed tail. Adults are blackish or dark gray, but faint yellowish crossbands are sometimes evident. The young are black with conspicuous yellow crossbands.

Alligators inhabit freshwater swamps, marshes, abandoned rice fields, ponds, lakes, and backwaters of large rivers in the lower Coastal Plain and a few portions of the upper Coastal Plain of the Carolinas. Their range once extended north to the southern margin of the Dismal Swamp. There are no historical records from Virginia, but population recovery in North Carolina, including several recent sightings north of Albemarle Sound, may bode well for the expansion of the alligator's range into southeastern Virginia at some point in the future.

Alligators are carnivorous, taking prey ranging from insects to large mammals. In June or July, a female builds a large mound of leaves, mud, and debris about 60 cm (24 in.) high and 120 to 200 cm (47 to 78 in.) wide, usually in a shaded area a few meters from the water. She deposits an average of about 35 eggs in a cavity atop the mound and remains nearby, guarding the eggs until they hatch in late summer or early fall. Hatchlings average about 240 mm (9.5 in.) long. The female may assist in freeing her young from the nest mound and will often remain with them for months or even years after hatching, defending them from potential predators. Even large adult alligators usually avoid humans; but they are potentially dangerous if provoked, and they should be left alone and never fed or otherwise encouraged to associate humans with food.

A few decades ago, alligator populations were decimated by commercial collecting. Federal protection under the Endangered Species Act in the early 1970s allowed recovery of many populations. The species is currently federally listed as Threatened by Similarity of Appearance (to the Endangered American Crocodile), as Threatened in North Carolina, and as a species In Need of Management in South Carolina.

Order Testudines (Chelonia)

TURTLES

Turtles are the most ancient of all living reptiles. They are evolutionarily conservative and have changed little since their origin early in the Triassic period some 250 million years ago. These unique vertebrates possess a shell—a protective structure composed of an upper part, called the carapace, and a lower part, the plastron. The shells of most species are bony and covered with keratinized scutes. The limb girdles are enclosed within the greatly expanded rib cage, a unique feature within the vertebrates. The feet are elephantine in highly terrestrial forms, webbed in the aquatic ones, or modified as flippers in those that live in the open seas. The jaws lack teeth and are covered by a horny beak. Adults of modern species range in carapace length from about 70 mm (2.75 in.) to 1.9 m (7 ft.).

Turtles are so different from other reptiles that many systematists advocate grouping them in a separate class, Chelonia. For the purposes of this book, we have taken the traditional approach and retained them as an order within the Reptilia.

Most turtles are omnivorous; but some are carnivorous, and a few are herbivorous. In some species, the juveniles are carnivorous and become herbivorous as adults. Most species court and mate in the fall as well as in the spring. All species are oviparous. The eggs are usually laid in a hole dug in the soil by the female using only her hind feet. Most species oviposit in late spring, but some also lay one or more times in the summer or at other times of the year. The number of eggs in a clutch varies from 1 to about 300, depending on the species. In Virginia and the Carolinas, hatching usually takes place in late summer or fall, and the young of several species may overwinter in the nest and emerge early the following spring.

Humans interact with turtles in many ways. We eat their flesh and eggs and prepare numerous products from their skins, shells, and bones. Live turtles, especially hatchlings, once constituted a lucrative part of the pet trade. Like many organisms, turtles are adversely affected by such human activities as drainage, pollution, land clearing, and strip mining. Without doubt, the automobile is one of the greatest threats to turtles, and each year many thousands are killed on the roads. Long lived and slow to mature, turtles are highly susceptible to such impacts on their populations. Several species are threatened with extinction, and while many are now protected by various laws, those laws are often easily circumvented, seldom enforced, and in most cases fail to address the greater threat of habitat destruction. One of the most serious current threats to turtles is a huge commercial food market in Southeast Asia. Turtle populations in that region have been decimated, and the continual demand has extended to include turtles worldwide. This "Asian Turtle Crisis" will probably result in the extinction of some species and will probably be eliminated or resolved only by means of a major shift in human values.

Most of the more than 290 modern turtle species occur in the warmer parts of the world, but at least 24 species are native to the Carolinas and Virginia. All but two of our species (Gopher Tortoise and Eastern Box Turtle) are chiefly or highly aquatic. The Emydidae includes at least 11 species and is the largest family.

Common Snapping Turtle
Chelydra serpentina

200 to 470 mm (8 to 18.5 in.) These large, brownish turtles are imposing and highly defensive. They attain weights up to 26 kg (57 lbs.) but can get much larger in captivity. Their most outstanding features include a large head; a long, tapering tail armed above with large scales; a small, cross-shaped plastron; and a carapace with 3 longitudinal keels most prominent in the young. In males, the cloaca lies posterior to the edge of the shell; in females, anterior.

Snapping Turtles are common in most freshwater habitats throughout the area. They occasionally enter brackish water. Underwater, they are usually inoffensive, but on land most individuals lunge and snap vigorously at imposing objects. Their powerful jaws and strong, heavily clawed limbs demand caution in handling. A wide array of invertebrates, small vertebrates, and plants are included in the highly varied diet.

In early spring, adults frequently wander from one body of water to another. In late spring or early summer, females lay 10 to 83 (usually about 25) eggs per clutch in a shallow nest, sometimes a considerable distance from water. The eggs are spherical and hatch in about 3 months; the young average about 30 mm (1.2 in.) in carapace length. The flesh of this turtle is a popular ingredient of soups and stews.

Striped Mud Turtle
Kinosternon baurii

75 to 120 mm (3 to 4.75 in.) This turtle was only recently discovered in our region, having long escaped detection due to its close resemblance to the Eastern Mud Turtle. It is fairly widespread and common farther south. Striped Mud Turtles have 3 longitudinal yellowish stripes on the carapace in the southern part of their range, but these markings are often faint or absent in individuals from our region. Most have pale yellowish stripes on the head and neck, similar to those of Eastern

Musk Turtles. A yellowish line is usually present on each side of the snout between the eye and nostril. These turtles are usually darker than Eastern Mud Turtles, with dark brown to black shells and dark gray or blackish skin.

Striped Mud Turtles are most often found in blackwater swamps, streams, and ponds, especially those dominated by cypress. They occur from southeastern Virginia southward throughout much of the Coastal Plain of the Carolinas and are also known from Franklin and Wake counties in the North Carolina Piedmont and from Bodie Island on the Outer Banks.

These turtles forage on the bottom for insects, snails, worms, and other invertebrates. Much remains to be learned about their natural history in our area. Females produce complements of from 3 to 7 elliptical, hard-shelled eggs about 27 mm (1.25 in.) long. Hatchlings are about 22 mm (0.9 in.) in carapace length.

Eastern Mud Turtle
Kinosternon subrubrum

75 to 124 mm (3 to 5 in.) This small turtle is appropriately named, reflecting both its color and habitat. Mud Turtles differ from Musk Turtles in that the pectoral scutes meet only narrowly on the midline of the plastron, which is hinged both before and after the abdominal scutes. This species differs from the Striped Mud Turtle by lacking distinct head stripes and a pale line between the eye and nostril.

This is an abundant turtle in the Coastal Plain and prefers the quiet waters of creeks, ditches, ponds, and lakes. It even tolerates brackish water and occurs on the Delmarva Peninsula and its barrier islands, as well as on the Outer Banks and other barrier islands in the Carolinas. It becomes increasingly scarce across the Piedmont and is absent from most of the Mountains.

Mud Turtles may be active both day and night. Insects, mollusks, crustaceans, carrion, and vegetation are included in their varied diet.

Most individuals leave their aquatic habitats in fall and overwinter in shallow burrows on land, a habit shared with the Chicken Turtle. Mating has been observed in late March. In late spring, a female deposits 2 to 6 elliptical, brittle-shelled eggs, usually in a shallow nest dug in soft soil, rotten wood, or surface litter near the water, but sometimes left uncovered on the surface. Hatching occurs about 100 days later, but the hatchlings, which average about 22 mm (0.9 in.) in carapace length, may remain in the nest until the next spring.

Stripe-necked Musk Turtle
Sternotherus minor

75 to 114 mm (3 to 4.5 in.) This small turtle has pale yellowish and dark stripes on the posterior head and neck, a relatively high brown or gray carapace with dark flecks or streaks, and a pinkish or yellowish, weakly hinged plastron. The carapace of juveniles usually bears a distinct middorsal keel and traces of an additional keel on each side. The shells of adults typically lack keels or are keeled only posteriorly. The vertebral scutes tend to overlap. There are

2 barbels on the chin but none on the throat. The Eastern Musk Turtle is similar but usually has prominent yellow or white stripes on the head and barbels on the chin and throat. When handled, most Musk Turtles release an unpleasant-smelling musk from glands along the inside margins of their shells.

Stripe-necked Musk Turtles are highly aquatic. They inhabit swamps, rivers, streams, and springs, preferring muddy bottoms near submerged logs or other objects. This species is known in our area only from extreme southwestern Virginia and from the Hiwassee and French Broad river systems in extreme western North Carolina.

This active turtle is diurnal and forages mainly during the morning hours. It is omnivorous but prefers aquatic insects and mollusks. The natural history of this species in our area is poorly known. It is listed as a species of Special Concern in North Carolina.

Eastern Musk Turtle
Sternotherus odoratus

80 to 136 mm (3 to 5.5 in.) This small, brownish turtle, also commonly called Stinkpot, is noted for its musky odor, a defense against predators. Its head bears 2 pairs of barbels on the chin and throat, a light yellowish line above the eye, and another below the eye. These lines may be obscure or lost in old males. The plastron is small, and its anterior lobe is weakly hinged. The suture between the humeral scutes (anterior pair) is about as long as that between the pectorals (second pair). In Mud Turtles, the pectoral suture is much shorter and the pectorals are triangular.

This turtle is abundant in ponds and streams in the Coastal Plain and lower Piedmont but is scarce in the upper Piedmont and in Mountain valleys. It prefers soft bottoms, where it will hibernate buried in the mud.

Eastern Musk Turtles are omnivorous, with insects and snails the most common foods. The 2 to 8 eggs are white, brittle, elliptical, and about 27 mm (1.1 in.) long. They are typically laid in soft dirt or humus, usually very close to the water. Some females produce more than 1 clutch in a season. Communal nests have been reported. The tiny hatchlings usually emerge in late summer or early fall. They are about 21 mm (0.8 in.) long and triangular in cross section.

Painted Turtle *Chrysemys picta*
114 to 180 mm (4.5 to 7 in.) Painted
Turtles in our area have pale transverse seams between the dorsal
scutes on a dark brown, greenish, or
black carapace. There are 2 yellow
spots in line behind the eye and conspicuous red markings around the
edges of the shell. The dark skin on
the legs is striped with red and yellow. Males are smaller than females
and have elongated nails on the forelimbs that are used to stroke the face
of the female during courtship.

Painted Turtles occur throughout
most of our area but are widely scattered in the Mountains and absent
from the Coastal Plain of South
Carolina and extreme southeastern
North Carolina. Popular as pets,
they have been introduced in many
localities. They prefer quiet waters
with muddy bottoms and plentiful
vegetation and are omnivorous feeders. Individuals commonly bask on
logs or stumps or float at the surface
of the water. Estimates of over 100
turtles per acre of water have been
reported.

Most nesting occurs in May and
June, when 3 to 9 eggs are laid in a
nest dug by the female, often in very
hard-packed soil. They require about
75 days to hatch. Some hatchlings
overwinter in the nest and emerge
the following spring. Hatchlings are
about 27 mm (1.1 in.) in carapace
length.

Spotted Turtle *Clemmys guttata*
89 to 127 mm (3.5 to 5 in.) These
small, blackish turtles are easily
recognized by the bright yellow or
orange spots on the head and carapace. The male usually has brown
eyes and a tan or pale brown chin;
the female, yellow or orange eyes
and a yellowish or pinkish chin.
Spots on the shells may sometimes
be few or occasionally absent, especially in small juveniles and old
adults. The plastron is orange, yellowish, or pinkish with large brown
or black patches.

This colorful species occurs
throughout the Coastal Plain and in
some sections of the lower Piedmont
in all three states. Scattered populations occur in the Shenandoah
Valley of Virginia. Favorite habitats
are damp meadows and pastures,
swamps, small streams, ditches,
woodland pools, and other shallow
bodies of water. Spotted Turtles
are especially vulnerable to habitat
disruption resulting from drainage,
timbering, and development.

These turtles are most active in
spring, and some individuals may be
abroad during warm periods in winter. They usually are difficult to find
during hotter parts of the summer.
Plants and small animals, especially
invertebrates, are the main foods. A
female deposits from 2 to 6 eggs in
late spring or summer, and hatching
takes place in late summer or fall.
Some hatchlings may overwinter in
the nest. Hatchlings average about
30 mm (1.2 in.) in carapace length.

The Spotted Turtle is listed as a
species In Need of Management in
South Carolina.

Wood Turtle
Clemmys insculpta (or *Glyptemys insculpta)*

140 to 230 mm (5.5 to 9 in.) This rare turtle is distinguished by its rough, sculptured carapace. Each dorsal scute is a raised pyramid of concentric ridges and increases in height with age. The tail is long, and the forelegs and necks of adults are usually marked below with orange. The yellow plastron has a prominent dark blotch on the posterior corner of each scute.

The Wood Turtle is a northern species that enters our area only in northern Virginia, south to Rockingham County. It has a restricted home range and is more or less terrestrial and diurnal. It forages through deciduous woods, wetlands, and fields in normal weather, but it moves to water in dry periods and to overwinter. Its varied diet includes algae, grasses, leaves, berries, insects, mollusks, earthworms, and tadpoles.

Wood Turtles nest from late May to July. A female digs a nest with her hind legs as deep as their length permits, wherein she deposits 4 to 17 eggs, each about 40 mm (1.5 in.) long. Hatchlings, 30 to 35 mm (1.3 in.) in carapace length, may emerge in the fall after about 2 to 3 months, but sometimes they overwinter in the nest.

The Wood Turtle is listed as a Threatened species in Virginia.

Bog Turtle

Clemmys muhlenbergii (or *Glyptemys muhlenbergii*)

76 to 102 mm (3 to 4 in.) This is our smallest turtle. The carapace is dark brown, and the plastron is variously marked with yellowish, brown, or black. The bright orange or yellow blotch on each side of the head and neck makes identification easy. The female has a shorter tail than the male and a flatter plastron with a wide notch at its posterior margin.

The Bog Turtle occurs in the Mountains and upper Piedmont of North Carolina and southwestern Virginia and is known from a very few sites in Greenville and Pickens counties, South Carolina. These secretive turtles are generally rare and occur in widely scattered, often isolated populations. They inhabit sedge meadows, marshy pastures, fens, and a variety of other mucky, usually open-canopied, spring-fed wetlands. When disturbed, they quickly burrow into mud or debris.

Bog Turtles are known to eat insects, worms, snails, slugs, amphibians, carrion, and seeds. Mating has been observed in April and May. In June or July, 3 to 5 eggs are laid in a shallow, often poorly concealed nest, usually in moss or a grass or sedge tussock. They hatch about 55 days later, and the young average about 25 mm (1 in.) in carapace length.

Northern populations of the Bog Turtle (those in the northeastern U.S.) are federally listed as Threatened; southern populations (those in our area) are listed as Threatened by Similarity of Appearance. The species is also listed as Endangered in Virginia, as Threatened in North Carolina, and as a species In Need of Management in South Carolina. Despite this protection, it faces illegal collecting for the pet trade and the much greater threat of habitat loss.

Chicken Turtle
Deirochelys reticularia

100 to 250 mm (4 to 10 in.) A very long neck and narrow shell identify this uncommon turtle. The fore-limb bears a unique broad yellow band, and the posterior surfaces of the thighs are barred vertically with yellow on black, a characteristic shared only with the Yellow-bellied Slider. The yellow plastron is often unmarked, but there are usually 1 or 2 elongate black blotches on the bridge. The dark carapace has a reticulate pattern of light lines. Females grow larger than males.

In our area, Chicken Turtles are known from two locations in south-eastern Virginia, from portions of the outer Coastal Plain and Sandhills of North Carolina, and from most

of the South Carolina Coastal Plain. They inhabit quiet water, avoiding rivers, and are especially partial to ponds, Carolina bays, ditches, and other shallow-water habitats in pine savannas. They spend considerable time on land, where they often over-winter. During droughts, they have been known to spend nearly 2 years inactive in shallow burrows in sandy uplands.

Nesting is prolonged, often occurring in fall, winter, or very early spring. Gravid females have been found nearly year-round. Five to 15 elliptical eggs are laid at a time, and there may be more than 1 clutch per year. Hatchlings average about 29 mm (1.1 in.) in carapace length and grow 25 to 30 mm a year until maturity—about 100 mm for males and 150 mm for females. These turtles are highly carnivorous—the young more so than the adults. Aquatic insects comprise much of the diet, but other invertebrates, small vertebrates, and some plants are also eaten.

Chicken Turtles are apparently in decline over much of their range. The species is listed as Endangered in Virginia.

Common Map Turtle
Graptemys geographica

Males 100 to 160 mm (4 to 6 in.), females 175 to 270 mm (7 to 10.5 in.) Common Map Turtles are olive green to brownish with narrow yellow stripes on the head and limbs and yellow reticulations on the carapace. There is a small yellow postorbital spot, and 1 yellow neck stripe usually turns upward across the tympanum. The plastron is yellow and unmarked in adults, but juveniles have dark markings following the sutures. The carapace has a low but distinct vertebral keel and is strongly serrate posteriorly. The crushing surfaces of the jaws are broad and smooth. Similar freshwater turtle species lack a distinct vertebral keel in adults, and the crushing surfaces of the jaws are ridged or toothed. Female Map Turtles are larger, with broader heads and heavier jaws than males.

The Common Map Turtle has been confirmed in our area only from the Tennessee River system in southwestern Virginia and from the Hiwassee System in Cherokee County, North Carolina. These turtles prefer large waters without strong currents. They often bask on rocks, logs, and snags but are wary and difficult to approach. The massive jaws of females are well adapted to crushing their favorite foods — mollusks and crayfish. The males probably eat more insects. In early summer, a female lays 10 to 16 eggs in a flask-shaped hole dug in the river bank. Hatchlings average about 32 mm (1.25 in.) in carapace length. Much remains to be learned about the distribution and natural history of this turtle in our area.

Diamondback Terrapin
Malaclemys terrapin

Males 100 to 140 mm (4 to 5.5 in.), females 150 to 230 mm (6 to 9 in.) This unusual turtle has a gray, brown, or black carapace and a lighter plastron of greenish to yellow. The skin of the neck and legs may be grayish with black spots or linear flecks, or completely dark. The individual carapacial scutes have prominent concentric age rings that may be worn smooth in old specimens.

This salt marsh turtle is found in the tidal channels of our sounds and estuaries that are bordered chiefly by *Spartina*. It is mostly carnivorous, preying heavily on mollusks (especially periwinkle snails), crustaceans, and other invertebrates.

The terrapin was the epicure's delight almost a century ago when it sold for nearly a dollar an inch to provide soup for the table. Populations were severely depleted before a slumping market and changing tastes permitted a marked recovery. Commercial collection is now illegal in much of our area, but at least some of these turtles are still taken illegally for the seemingly insatiable Asian turtle market. Crab pots, especially abandoned ones, are also a serious threat to populations throughout the range.

Females seek suitable nest sites above the high-tide mark and deposit complements of 2 to 12 oval-shaped eggs, averaging about 30 mm (1.2 in.) long. A mature female may lay 2 clutches per season beginning in May or June. Hatching occurs in August or September, but hatchlings may remain in the nest and emerge the following spring. They average about 29 mm (1.1 in.) in carapace length.

The Diamondback Terrapin is protected as a species of Special Concern in North Carolina.

River Cooter *Pseudemys concinna*
230 to 333 mm (9 to 13 in.) This
large basking turtle occurs in two
highly variable forms that have
been variously regarded as separate species, as subspecies, and as
ecological variants. An essentially
riverine Piedmont form usually has
a flatter shell with a light concentric
pattern that includes a rear-facing
C-shaped mark on the first and/or
second pleural scute on each side,
and a reddish plastron with dark
markings along the seams between
the scutes. The form more common
in quiet Coastal Plain backwaters,
swamps, canals, and ponds has a
higher shell with transverse light
marks that branch irregularly, and
usually a plain yellowish plastron.
This form is sometimes considered
a separate species (*P. floridana*,
the Florida Cooter). In our area the
two show little genetic variation
and freely interbreed, with many
specimens exhibiting intermediate
characteristics. The heads, necks,
and limbs of both forms are striped
with yellow, and the hind limbs have
both vertical and horizontal yellow
markings posteriorly. The usually
smaller males have long, needlelike
foreclaws, used to stroke the head of
the female during courtship.

This species complex occurs from
central and southeastern Virginia
south through most of the Coastal
Plain and Piedmont of the Carolinas.
It is absent from the Mountains and
much of the western Piedmont.

Female Cooters lay from 5 to 20
elliptical eggs per clutch, usually
in May or June. The eggs average
about 35 mm (1.4 in.) long. Nests are
usually dug in soil in open areas near
water. Hatching takes place in about
90 days. Emergence from late nests
may be delayed until the next spring.
Adults are largely herbivorous, feeding on a variety of aquatic plants.
The young are more omnivorous.

The ecological and genetic relationships between turtles in this
variable and confusing species complex deserve careful study.

Red-bellied Cooter
Pseudemys rubriventris

250 to 400 mm (10 to 16 in.) This large basking turtle has a notch on the midline of the upper jaw flanked by a downward-projecting cusp on either side, and both jaws are strongly serrate on the cutting edges. The plastron is reddish and usually bears dark smudges or a symmetrical marking. The marginal and pleural scutes have dull red vertical bars through their centers; these markings may be obscure in old individuals. The top of the head usually has an arrow formed by a median light stripe joining with a line from the eye to the tip of the snout on each side. The closely related River Cooter lacks the reddish tones and the arrow on the head, and its upper jaw edges are nearly smooth.

The Red-bellied Cooter occurs from rivers in the Shenandoah Valley and the Potomac River in Virginia south through the Coastal Plain to Pamlico Sound in North Carolina. Hybridization with the members of the River Cooter complex has been reported from the Albemarle Peninsula in North Carolina. Egg complements of 9 to 35 are laid in suitable soil near the water in May to July. Some females may lay 2 clutches in a season. Young average 33 mm (1.3 in.) in carapace length and may emerge in late summer or overwinter in the nest until the following spring.

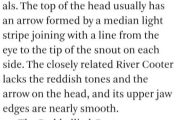

Eastern Box Turtle
Terrapene carolina

114 to 165 mm (4.5 to 6.6 in.) This familiar turtle has a brown, dome-shaped carapace, variously mottled with yellow or orange, and a hinged plastron that allows the turtle to withdraw and enclose its head, limbs, and tail within the shell. Adult males have concave plastrons and often red eyes; adult females have flat plastrons and usually brown or yellow eyes.

Box Turtles live in forested habitats throughout the area up to about 1,300 m (4,260 ft.) elevation.

These turtles are largely terrestrial, but they often enter water during hot, dry weather. In the early morning and after rains, individuals frequently wander across roads, where many are killed by cars. Box Turtles eat a wide variety of plants and small animals. Several kinds of highly toxic mushrooms are included in their diet, and cases of poisoning are known in some persons who have eaten these reptiles. A female lays from 2 to 7 soft-shelled, elliptical eggs, usually in June or July, and hatching occurs in about 3 months. Hatchlings average about 30 mm (1.2 in.) in carapace length. Some may overwinter in the nest.

The Eastern Box Turtle is the official state reptile of North Carolina. Habitat loss and fragmentation have precipitated population declines in this once abundant turtle.

Yellow-bellied Slider

Yellow-bellied Slider
Trachemys scripta

125 to 290 mm (5 to 11.5 in.) Young Yellow-bellied Sliders have green or olive shells with prominent yellow transverse bars on each side. A large, oblique yellow blotch behind the eye linking a pair of yellow stripes on each side of the neck is diagnostic of this turtle. The plastron is yellow (but sometimes stained brown or reddish), usually with a pair of dark spots on the gular (anterior) scutes and sometimes with additional spots on more posterior scutes. The

posterior thighs are barred vertically with yellow and black. These colors and patterns often become obscure with age, and some older individuals may be nearly black. Adult males are smaller than females and have elongated nails on the forelimbs.

This turtle is abundant in the Coastal Plain and portions of the Piedmont north to the middle peninsula of Virginia. It also occurs in parts of southwestern Virginia and in the French Broad River in Madison County, North Carolina. Turtles in these montane populations are known as Cumberland Sliders (*T. s. troostii*) and differ in usually having a dark smudge on each plastral scute and a less prominent, often broken, yellow patch behind the eye.

These aquatic turtles often bask, especially in early spring, and their common name derives from their habit of quickly "sliding" into the water when disturbed. Nesting usually occurs in May and June. Females leave the water to find

Cumberland Slider (profile)

Red-eared Slider

suitable sites for nests, which they excavate with their hind legs. Four to 12 eggs are laid, which hatch in about 2 months. Some hatchlings overwinter in the nest, emerging the following spring. This species is omnivorous, but juveniles are more carnivorous than adults.

A third subspecies of this turtle, the Red-eared Slider (*T. s. elegans*), has unfortunately been widely introduced via the pet trade. It has become established in some portions of our area, where it frequently interbreeds with the native Yellow-bellied Sliders.

Gopher Tortoise
Gopherus polyphemus

150 to 368 mm (6 to 14.5 in.) This large terrestrial turtle has a high, dorsally flattened brown or tan carapace. The carapacial scutes have light centers, which are yellow or orange in the young. There are 2 large anterior (gular) projections on the unhinged, dull yellow plastron, a feature not present in other turtles of the area. The head, legs, and tail are brown or gray. The front legs are shovellike and heavily scaled; the hind legs are elephantine in shape.

Gopher Tortoises live in rolling sandhill areas normally characterized by wide patches of bare sand, scattered clumps of Wiregrass, Turkey Oak, and Longleaf Pine. Native, reproducing populations of this rare species are known in our area only from Jasper, Hampton, and Aiken counties, South Carolina.

This turtle excavates an extensive burrow up to 10 m (30 ft.) long; the entrance is usually marked by a small mound or "apron" of sand. It is diurnal and feeds mainly on grasses, fruits, and other plant material. In late spring or early summer, females deposit 4 to 7 spherical eggs in nests usually located far from the burrow. Many other animals may occupy tortoise burrows, including Gopher Frogs and Eastern Diamondback Rattlesnakes.

The Gopher Tortoise is protected as an Endangered species in South Carolina.

Leatherback Sea Turtle
Dermochelys coriacea
1,180 to 1,890 mm (46 to 74 in.); 295 to 910 kg (650 to 2,000 lbs.) This unmistakable monster among turtles is the heaviest living reptile. Instead of hard scutes, its shell consists of leathery skin covering a mosaic of small irregular bones. Its carapace has 7 longitudinal ridges; its plastron, 5. There are no claws on the flippers. The color is dark brown or black, sometimes with light blotches on the head, neck, and flippers, especially in the young. The neck is short and not retractile as in other sea turtles.

Although primarily tropical, the Leatherback wanders farther north than other sea turtles and may reach Nova Scotia and Newfoundland. It is capable of maintaining a deep body temperature far above that of the water, apparently through muscular activity and fatty insulation. A powerful swimmer, it is the most pelagic of all turtles and the most wide ranging of all reptiles. Leatherbacks are omnivorous. They feed extensively on jellyfish, including the Portuguese man-of-war, but also eat sea urchins, crustaceans, fish, mollusks, and plants.

Leatherbacks nest at night on widely scattered tropical beaches, often sharing them with other species of marine turtles. A few individuals have nested on North Carolina barrier islands. A female will lay several times a season at about 10-day intervals, but most females nest only every 2 to 3 years. Usually about one-third of the 50 to 170 spherical eggs laid are yolkless. Hatchlings average about 63 mm (2.5 in.) in carapace length.

This species is everywhere rare. As with all sea turtles, the eggs are edible and the nests are often raided by humans and other predators. The flesh, however, is nowhere esteemed, but it does provide an oil used commercially as a base for many cosmetics.

The Leatherback Sea Turtle is federally listed as an Endangered species.

Loggerhead Sea Turtle
Caretta caretta

800 to 1,200 mm (31 to 47 in.); 77 to 227 kg (170 to 500 lbs.) This is the sea turtle most often seen on our coast and in the Chesapeake Bay. Large with a massive head, it is the only reddish brown species. The nuchal scute contacts the first of 5 pairs of pleural scutes, and usually 3 scutes on each side form the bridge between the carapace and plastron.

The Loggerhead is basically tropical and subtropical, but it has been recorded from Argentina to Newfoundland and the British Isles. These animals are great wanderers, and tagged individuals have been taken well over 1,600 km (1,000 mi.) from the point of marking. They are largely carnivorous, feeding on fish, crustaceans, mollusks, and a variety of other marine animals.

Loggerheads nest sparingly at least as far north as Chincoteague Island, Virginia, and commonly in the Carolinas. Like most sea turtles, they nest on isolated beaches in late spring and early summer. Mating occurs just beyond the surf. At night the female crawls ashore and selects a suitable site for her nest, usually along the dune front. With her hind flippers, she excavates an egg chamber 15 to 25 cm (6 to 10 in.) in diameter and 12 to 50 cm (5 to 20 in.) below the surface. Into this she lays from 52 to 341 spherical eggs. She then covers the nest by scraping sand and crawling over it until the precise location is obscured. An individual lays on a 2- or 3-year cycle but will deposit 2 to 6 clutches at about 2-week intervals in her nesting year. The hatchlings, averaging about 45 mm (1.75 in.) in carapace length, emerge at night about 2 months later and make their way to the sea. Illumination, slope of the beach, and the white surf probably are directional cues to the young. Many eggs and hatchlings are eaten by raccoons, foxes, ghost crabs, and other predators, but disorientation due to lights from beach houses is probably a much greater threat. Despite intensive conservation and education efforts on behalf of sea turtles, many human activities threaten their continued survival.

The Loggerhead is federally listed as a Threatened species. It is the official state reptile of South Carolina.

Green Sea Turtle
Chelonia mydas

760 to 1,530 mm (30 to 60 in.); 100 to 295 kg (220 to 650 lbs.) This large sea turtle is distinguished from the others by a single pair of elongate scales between the eyes, a strongly serrate lower jaw, and a single claw on each front flipper. The Green and the Hawksbill have 4 instead of 5 pairs of pleural scutes.

Green Sea Turtles are pantropical in distribution. They readily migrate long distances across open seas but spend most time in shallow waters, feeding mainly on eelgrass (*Zostera*), mangrove, turtle grass (*Thallasia*), and other plants. This turtle is rare in our area, but it nests on North and South Carolina beaches occasionally. Females nest at night, depositing from 3 to 9 clutches of 75 to 200 large, spherical eggs each, at about 2-week intervals between June and early September. Females normally nest only every second, third, or fourth year. Hatchlings typically emerge at night from August to October and average about 50 mm (2 in.) in straight-line carapace length.

This species is well adapted for prolonged diving and can survive up to 5 hours with no detectable oxygen in its blood and with 9-minute intervals between heartbeats.

The common name is derived from the color of the body fat, which lends to its meat an exquisite flavor. The Green Turtle has been termed the most commercially valuable reptile in the world. It once occurred in large numbers in the Caribbean region; however, many breeding colonies have been greatly reduced, and some have been exterminated. Because the eggs and flesh of this species are eagerly sought, the future for it is dim. Many nesting colonies are finally being protected, and the weaker ones are receiving egg transplants. The intensive studies of reproduction and migration that are under way should provide a basis for sustained management. Adequate international cooperation must be realized.

The Green Sea Turtle is federally listed as a Threatened species.

Hawksbill Sea Turtle
Eretmochelys imbricata

750 to 915 mm (29.5 to 36 in.); 36.5 to 127 kg (80 to 280 lbs.) The large scutes of the carapace of this small sea turtle overlap (except in the very young and old); hence, the species' scientific name. The head is relatively small with a narrow snout and smooth cutting edges on the jaws, suggesting a hawk's bill. The carapace, like that of the Green Turtle, has 4 pairs of pleural scutes, the first of which does not contact the nuchal. The Hawksbill has 2 pairs of scales between the eyes and nostrils, unlike the Green Turtle, which has but 1 pair.

The Hawksbill inhabits tropical seas and only rarely wanders into our area, where it has not been reported to nest. It prefers coral reefs and rocky ledges, where it forages for a wide variety of sessile and sluggish invertebrates—especially sponges. It also consumes algae, mangroves, and many other types of plants.

A Hawksbill is very defensive when molested and can inflict painful bites. This turtle was the traditional source of "tortoiseshell," a product used in many ornamental articles, and it has been widely exploited for that and other reasons. Modern plastics alleviated some pressures on this turtle, but the demand for genuine tortoiseshell remains, and the mostly illegal trade in these turtles continues.

Hawksbills nest earlier and have more diffuse breeding patterns than the highly gregarious Green Turtle. This behavior makes them and their nests more difficult to protect against predation. Females probably nest every 2 or 3 years and lay 2 or more clutches of about 160 eggs at about 2-week intervals. Mating occurs immediately after a spent female returns to the sea. Like the other marine turtles, little is known about the young once they enter the sea.

The Hawksbill is federally listed as an Endangered species.

Kemp's Ridley Sea Turtle
Lepidochelys kempii

580 to 750 mm (23 to 30 in.); 36 to 50 kg (79 to 110 lbs.) This smallest of our sea turtles has a heart-shaped carapace about as broad as it is long. Among sea turtles the Ridley is unique in its gray coloration and in having a secretory pore near the posterior edge of each scute forming the bridge. The nuchal scute contacts the first of 5 pairs of pleural scutes as in the Loggerhead.

Kemp's Ridley inhabits the Gulf of Mexico and the East Coast, straying to Nova Scotia and Europe. It is uncommon along our shores, but several enter the Pamlico and Albemarle sounds and Chesapeake Bay to forage annually. This turtle feeds largely on crabs and mollusks, but it also eats other invertebrates and plants. Captives are nervous and short tempered.

The Ridley differs from other sea turtles in its extraordinary communal, diurnal breeding effort. Immense groups, called *arribadas*, of as many as 40,000 females once came ashore on a stretch of beach about 2 km (1.25 mi.) long in Tamaulipas near Rancho Nuevo, Mexico, on a single day. Many of these turtles have been killed in shrimp nets, and today only about 800 to 1,000 turtles nest each year. There are usually 3 such *arribadas* in a year, at slightly different sites, about 10 days apart, between late March and the end of June. As many as 180 eggs per clutch are laid in the first nesting, and as few as 80 in the third. Nesting normally takes 4 to 6 hours. Coyotes, feral dogs, and other predators gather in great numbers during these nestings and destroy most eggs. Vultures and predatory fishes account for the loss of many hatchlings. Humans were until recently major predators on both eggs and adults, but the nesting beaches are now protected.

This remarkable breeding pattern and its site were discovered only a few decades ago. Individual Ridleys occasionally nest elsewhere, and a very few individual nestings have been documented on North and South Carolina beaches.

Kemp's Ridley is one of the most imperiled reptiles in the world. It is federally listed as an Endangered species.

Florida Softshell *Apalone ferox*
Males 150 to 292 mm (6 to 11 in.),
females 200 to 498 mm (8 to 19.5 in.)
This large, pancake-shaped turtle
has a very long neck; a narrow head
with a slender, snorkellike snout;
and a soft, leathery shell that lacks
horny scutes. The dark brown to
brownish gray carapace has a distinct marginal ridge, small hemi-
spherical anterior projections, and
in some individuals, large, obscure
dark spots. The feet are extensively
webbed. Spiny Softshells lack both
the small projections on the anterior margin of the carapace and
the marginal ridge, and their skin is
more heavily patterned. All softshells
should be handled with extreme caution; their sharp claws and strong
jaws are capable of inflicting serious
injury.

In our area, the Florida Softshell
lives in rivers, ponds, and lakes only
in the Combahee and Savannah river
systems in southern South Carolina.

Females deposit 10 to 22 eggs in
early summer in sunlit, sandy areas
on the banks of rivers. Incubation requires about 2½ months. This turtle
is omnivorous, but prefers crayfish,
mollusks, frogs, and fish.

Spiny Softshell *Apalone spinifera*
Males 125 to 235 mm (5 to 9 in.),
females 180 to 450 mm (7 to 18 in.)
The leathery carapace and tubular
nostrils of softshells are so distinctive that they cannot be confused
with any hard-shelled turtle. This
species differs from the Florida Softshell by having spines or tubercles
on the front edge of the carapace,
stripes on the head and neck, and
strongly patterned feet. Juveniles
and males have sharp black spots, or
ocelli, on the back; adult females, indistinct blotches. The large nostrils
have a ridge on each median side.

Two subspecies of this turtle
occur in our area. The Eastern Spiny
Softshell (*A. s. spinifera*) is known
primarily from the Clinch and Holston systems in southwestern Virginia and the French Broad system
in North Carolina. The Gulf Coast
Spiny Softshell (*A. s. aspera*) has
been found in all Atlantic drainage
systems south and west of the Cape
Fear.

These turtles occur in large

streams with sandy bottoms and in
ponds and lakes. They often bury
themselves in shallow water and
extend their long necks to permit
the terminal nostrils to break the
surface for breathing. The eggs are
spherical and about 22 mm (0.9 in.)
in diameter. They are deposited 10
to 25 cm (4 to 10 in.) deep in sandy
banks.

Much remains to be learned
about the distribution, status, and
natural history of this turtle in our
area. The Eastern Spiny Softshell is
listed as a taxon of Special Concern
in North Carolina.

Order Squamata

The Squamata are the most numerous of living reptiles. They originated in the early Mesozoic from lizardlike forms and diversified in late Mesozoic and Cenozoic times. They have many salient traits: (1) The lower jaw joins a movable bone, the quadrate, which in turn articulates with the skull. (2) The roof of the mouth bears an interesting chemoreceptor, Jacobson's organ. This chemosensory organ also occurs in amphibians and some mammals but is found only in the embryos of other reptiles. Snakes and many lizards often flick their long, narrow tongues out of the mouth, then upon retraction insert the tips into Jacobson's organ, which detects odorous particles adhering to the tongue. (3) The cloacal opening is a transverse slit rather than a longitudinal one as in other living reptiles. (4) This transverse opening is, in part, an adaptation for the unusual paired copulatory organs known as hemipenes. Either hemipenis can become engorged with blood and everted into the cloaca of the female. (5) The Squamata produce large eggs, but unlike other reptiles, their eggs lack albumin. Most are oviparous: some species retain the eggs in the distal part of the oviduct and lay them shortly before hatching; others retain the unshelled eggs until they hatch, and the young are born alive. In many live-bearing species, the metabolic exchange between the female and her young is completely respiratory, but in some there is a nutritional exchange as well. (6) Emergence of young squamates from the egg is facilitated by an "egg tooth." This true tooth erupts from the upper gum shortly before hatching, projects from under the lip, and cuts the egg membrane or shell as the young moves its head about. It is lost a day or two after birth. The egg tooth differs greatly from the broad, cornified "egg caruncle" of other reptiles, birds, and egg-laying mammals.

Suborder Sauria

This is the most successful group of living reptiles; about 3,000 species are known. Most lizards have 4 legs with 5 toes on each foot, but some (especially the burrowing forms) have lost toes and even legs. In contrast to snakes, most lizards have movable eyelids and a middle ear, a tympanum, and an ear canal. The quadrate bone is only slightly movable, and the right and left halves of the lower jaw are firmly united, thus restricting the size of the mouth opening and the size of prey eaten. In many lizards, the tail is easily broken and soon regenerates. Many have an armor of bony scutes (osteoderms) in the dermal layer of the skin under the scales. Adults of our native species range in total length from 80 mm (3 in.) (Ground Skink) to 1,180 mm (46.5 in.) (Slender Glass Lizard).

Lizards have undergone extensive adaptive radiation and have invaded a wide variety of habitats. Some are highly aquatic, and others are well adapted to life in the desert. Most are carnivores, but a few are vegetarians.

In spite of the great diversity and wide distribution of lizards, our fauna is strikingly impoverished. Only 14 species occur in the area, and 2 of these, the Mediterranean Gecko and the Texas Horned Lizard, have been introduced and are restricted to a few widely scattered populations. As might be expected, the warmest part of our area, the Coastal Plain of South Carolina, has the most species (13), and the Mountain region of Virginia has the fewest (6). Most native species are skinks (5) or glass lizards (4). None of our lizards is venomous, and all lay eggs and feed mainly on insects.

Mediterranean Gecko
Hemidactylus turcicus

100 to 127 mm (4 to 5 in.) This introduced species is tan, pale gray, yellowish, pinkish, or whitish with scattered light and dark spots and is unlikely to be mistaken for any of our native lizards. The pale belly and sides are somewhat translucent. The body, head, limbs, and tail are covered with numerous wartlike tubercles. The fragile tail is marked with dark bands; these are most prominent in juveniles. The toes are expanded, with broad, brushlike

pads on their undersurfaces. The large, protuberant eyes have vertical, elliptical pupils and transparent, immovable eyelids like those of snakes.

These geckos are native to the Mediterranean region of southern Europe, northern Africa, and southwestern Asia. Established populations have been reported in our area from Charleston, South Carolina; Wilmington, North Carolina; and Richmond, Virginia. This adaptable species and its eggs are often inadvertently transported about by humans. Now common in many parts of the southern U.S. and Mexico, it has rapidly expanded its range in recent years and may be expected to continue to do so.

Almost exclusively nocturnal, Mediterranean Geckos are usually seen around buildings, especially near lights at night, where they forage for insects and other arthropods. By day, they hide in wall crevices or other sheltered spots and may survive extremely cold weather by entering heated buildings. Numerous microscopic, adhesive bristles on their toe pads enable them to easily traverse walls and ceilings. Our only lizards with voices, they sometimes emit faint, mouselike squeaks. Dominant males may also utter barks or clicks. Females usually deposit a pair of brittle-shelled eggs 2 or 3 times a year, in crevices, under bark, or in moist soil. Communal nesting often occurs. Hatchlings are about 20 to 30 mm (0.75 to 1.25 in.) long.

Green Anole *Anolis carolinensis*
130 to 200 mm (5 to 8 in.) This highly arboreal lizard is familiar to everyone spending much time outdoors in the Coastal Plain of the Carolinas. In bright light and warm situations, an anole is usually a light emerald green, but under damp, cool, or dark conditions, a dull olive, brown, or gray color appears. The causative factors behind these color changes are complex. No other lizard in our area can undergo such color changes. Many anole species occur in the American tropics.

The Green Anole ranges throughout the Coastal Plain and southern Piedmont of North Carolina and all of South Carolina. It does not occur in Virginia. It is common in disturbed areas such as roadsides, forest edges, and old building sites with an abundance of shrubbery and sunlight.

The reproductive biology of this species has been studied intensely. The reproductive organs are regulated primarily by photoperiod. They enlarge during the warm, lengthening days of spring and regress with the shortening days of late summer. Males are strongly territorial. When one is approached by another male, he compresses his body, extends his bright pink dewlap, or throat fan, and bobs his head. If

the intruder continues to approach, aggressive fighting may ensue. Although a male's territory may be relatively small—as small as 1 cubic meter—it often encompasses the home ranges of 2 or 3 females. As in the human female, only 1 ovarian follicle develops at a time, and the ovaries alternate in egg production. A female anole is capable of producing 1 egg every 2 weeks for the entire breeding season, but before producing each egg she must be courted. The sight of a displaying male triggers ovarian development, which, in turn, evokes sexual receptivity and, later, ovulation. The female may store viable sperm for at least 8 months after copulation. Eggs are soft shelled and are laid in shallow depressions in moist soil, leaf litter, rotting wood, or tree cavities. They hatch in about 7 weeks, and hatchlings are about 65 mm (2.5 in.) long. Insects and other arthropods are the primary foods; but nectar may also be eaten, and these lizards have occasionally been known to visit hummingbird feeders.

Texas Horned Lizard

Phrynosoma cornutum

60 to 181 mm (2.5 to 7 in.) Texas Horned Lizards, erroneously called "horned toads," have dorsoventrally flattened, light brown or gray bodies. The back is covered with short spines and prominent dark spots. A 2-rowed fringe of triangular scales borders each side. Two stout spines on the top of the head project upward.

The native range of this introduced species includes Texas, Oklahoma, most of Kansas, and portions of adjoining states. Established populations occur on several beachfronts and barrier islands in Charleston, Colleton, Georgetown, and Horry counties, South Carolina, and in a residential area near Swansboro in Onslow County, North Carolina. These colonies occupy primarily open grassy or sandy areas.

Horned Lizards are diurnal.

Active mainly during the summer, they feed largely on ants—especially harvester ants. In the western United States, females deposit 23 to 37 eggs in a pocket of sand about 15 cm (6 in.) deep; these require 39 to 47 days to hatch. Members of this species can eject blood from the corners of their eyes to a distance of a few meters. The function of this bizarre habit is not fully understood, but it is thought to effectively repel certain predators.

Eastern Fence Lizard
Sceloporus undulatus

100 to 184 mm (4 to 7 in.) The Fence Lizard is gray or brown above with indistinct, darker, wavelike markings. Mature males have a bright greenish blue area bordered medially with black on each side of the belly. Conspicuous keeled scales give a rough appearance.

This abundant lizard occurs throughout most of our area. It inhabits open situations with plenty of sunlight, such as building sites, slab piles, open pine woods, fences, and rocky places. It avoids dense woods, is scarce at high elevations, and is absent from the Outer Banks and some other portions of the outer Coastal Plain. Fence lizards are diurnal. Active and alert, they frequently ascend trees to escape predators but are equally at home on the ground.

Emergence from hibernation occurs with warm, sunny weather in March. Courting males are territorial and defend against rival males by bobbing their heads and standing high to display their bright belly colors. Mating has been observed in April. Females lay from 5 to 16 eggs in late spring in burrows under rotten logs, mulch piles, or similar places, and hatching takes place in midsummer. A second clutch may be produced in an extended season. Hatchlings average about 50 mm (2 in.) long. Food consists mostly of insects and other arthropods.

Six-lined Racerunner
Cnemidophorus sexlineatus (or
Aspidoscelis sexlineata)
150 to 240 mm (6 to 9.5 in.) This lizard has 6 whitish to yellow longitudinal stripes on a dark brown to black
background. The dorsal scales are
tiny granules, but the belly scales are
large, quadrangular, and in 8 regular
rows. The tail is long and slim and
is brownish or gray in the adult but
bluish in the young. Racerunners
superficially resemble skinks, which,
however, are shiny and have brighter
blue tails as juveniles.

Racerunners occur from the
Outer Banks and other barrier
islands to the Mountains but normally avoid most elevations above
about 650 m (2,130 ft.). They occur
most commonly in the Coastal Plain
but sporadically in the Piedmont and
Mountains where hot, dry habitats
occur. Several populations have been
documented in the Shenandoah Valley. These alert inhabitants of open,
usually sandy areas can quickly take
cover under clumps of grass or in
underground burrows. Our fastest
terrestrial reptiles (their speed sur-
passed perhaps only by that of swimming Leatherback Sea Turtles), they
are often active in extremely warm
weather, when they are most difficult
to catch by hand.

Females deposit from 1 to 6 eggs
a few centimeters beneath the surface in sandy soil, decaying wood, or
similar substrate. The nesting season is prolonged, and older females
lay 2 clutches per season. The
hatchlings, averaging about 95 mm
(3.75 in.) in length, appear from late
June into September. Racerunners
forage widely in search of the insects
and other arthropods that constitute
their primary prey.

Coal Skink

Eumeces anthracinus (or Plestiodon anthracinus)

130 to 178 cm (5 to 7 in.) This small skink has 4 dorsal light stripes, whereas other *Eumeces* have 5 or 7. The light dorsal head stripes of other skinks (except older adults) are also absent. Coal Skinks have 1 scale behind the scale under the chin; other skinks have 2. Adults are olive gray to olive brown above and bluish or gray below. Virginia Coal Skinks have a light stripe through the posterior upper labials; the blue-tailed young

otherwise resemble the adults. Carolina Coal Skinks have light spots on the centers of these scales, and the blue-tailed young have black bodies.

Coal Skinks live under logs, rocks, or leaf litter on wooded, rocky hillsides, frequently near water. Widely disjunct populations are known in the area from the Mountains of Virginia and North Carolina and from Sassafras Mountain, South Carolina.

This lizard is uncommon and secretive, and much remains to be learned about its natural history in our area. More terrestrial than other *Eumeces*, it seldom climbs but readily enters water when pursued. Insects and other arthropods are the primary foods. A female deposits a clutch of 4 to 9 eggs in late spring or early summer and attends them until they hatch in late summer. Hatchlings are about 55 mm (2.25 in.) long.

The Coal Skink is listed as a species In Need of Management in South Carolina.

Five-lined Skink (young adult)

Five-lined Skink
*Eumeces fasciatus (*or *Plestiodon fasciatus)*
130 to 205 mm (5 to 8 in.) Our 3 large blue-tailed skink species are difficult to distinguish, especially in the field. The tail is brownish in adults but brilliant blue in juveniles. This common, familiar species has 5 light stripes on a dark background, whereas the other 2 species in our area usually have 7. In large males, the stripes are lost, and the head is coppery red throughout the breeding season. The lateral dark stripe on the tail is longest in this species, extending for more than half the tail's length. The Five-lined Skink is slightly smaller than the Southeastern Five-lined Skink, and its median subcaudal scale row is wider than the adjacent rows. It is considerably smaller than the Broad-headed Skink, averages 2 fewer rows

of scales around the body (28 to 30), and usually has 4 instead of 5 scales on the upper lip before the scale under the eye.

This species is absent in our area only from the Outer Banks and the higher elevations in the Mountains. Its total range corresponds closely with the eastern deciduous forest. Although it basks in sunlight on logs, fences, rocks, and buildings, it

Five-lined Skink (juvenile)

prefers a more mesic habitat than the other large skinks.

Skinks do not defend defined areas; hence, they are not truly territorial. But during the mating season in April and May, males are hostile to one another, and chance encounters result in physical combat, sometimes with serious injury.

A female lays from 2 to 15 eggs in rotten wood or under a sheltering object. The number of eggs varies according to her size. She guards the nest, turning the eggs daily, and will void on them if they become too dry. Eggs are usually laid in June or July, and hatching occurs about a month later. Hatchlings average about 65 mm (2.5 in.) in length. Sexual maturity is attained in the second spring. Skinks feed predominantly on arthropods and other invertebrates, often with a preference for large items such as spiders, crickets, grasshoppers, beetles, harvestmen, earthworms, and snails.

Southeastern Five-lined Skink
Eumeces inexpectatus (or Plestiodon inexpectatus)

140 to 216 mm (5.5 to 8.5 in.) This species is best distinguished from the other blue-tailed skinks by the narrow midventral scale row on the underside of the tail. Near the base of the tail, this row is little if any wider than the adjacent rows. The dorsolateral stripes are on the fourth and fifth lateral rows rather than the third and fourth as in the other species. This and the Broad-headed Skink in our area usually have a pair of faint stripes on the sides near the belly—7 light stripes in all. In juveniles, the head stripes and the mid-dorsal stripe usually do not meet, whereas in the other 2 species they converge at the back of the head. The snout and head stripes of hatchlings are orange.

The Southeastern Five-lined Skink is abundant in the Coastal Plain but occurs sparingly elsewhere in our area. It is the only large skink on the Outer Banks and on most coastal islands.

Common in pine flatwoods and sandhills, this skink also thrives in disturbed habitats such as recently lumbered lands, old house sites, and beach areas. Its habits are similar to those of the Five-lined Skink, but it is somewhat less arboreal. Insects and other arthropods are the primary foods. A female deposits a clutch of from 3 to 10 eggs under a sheltering object, usually in June or July, and attends them until they hatch several weeks later. Hatchlings are about 70 mm (2.75 in.) long.

Broad-headed Skink (adult male)

Broad-headed Skink
Eumeces laticeps (or Plestiodon laticeps)

165 to 325 mm (6.5 to 13 in.) This is our largest skink. It has 30 or 32 scale rows around the body and 5 upper labial scales anterior to the scale just beneath the eye. Large males lose all traces of stripes and have a broad head, which is red during the breeding season. Females are smaller and retain at least remnants of the striped pattern common to the blue-tailed skinks. This species differs from the Southeastern Five-lined Skink by having wide scales in the middle row under the tail. Its young differ from those of the Five-lined Skink by being larger and usually having 7 light stripes and longer toes and claws.

The Broad-headed Skink inhabits most of our area except the higher mountains and the Outer Banks. It occurs on Roanoke Island and on some barrier islands in South Carolina and is most abundant on some of the coastal estates and old plantations, where it finds its favorite habitat—large, spreading trees such as Live Oak, Water Oak, and cypress. It is highly arboreal and frequents living and dead trees to considerable heights. Throughout its range it prefers warmer and more xeric habitats than does the Five-lined Skink.

Skinks are known as "scorpions" to many rural folk, who erroneously consider them venomous. Because of its size, the Broad-headed Skink figures most prominently in these notions. A large one is capable of

Broad-headed Skink (adult female)

giving a painful but harmless nip to the unwary collector.

The 6 to 15 eggs, laid in June or July and attended by the female, usually hatch by September. Hatchlings are about 75 mm (3 in.) long. The habits of the Broad-headed Skink are similar to those of the Five-lined Skink; however, by being more arboreal, it sometimes nests in tree cavities and may select different food items.

Ground Skink *Scincella lateralis*
80 to 130 mm (3 to 5 in.) This diminutive skink is our smallest
reptile. It is shiny brown or bronze
with a dark stripe on each side, the
lower edge of which is ill defined
and blends into the whitish or yellow
belly. A transparent disk is present in
the lower eyelid. This lizard may be
mistaken at a glance for one of the
Dwarf or Two-lined Salamanders.

This species is common in most
of our area except in the Mountains,
where it is absent or local. Its near-

est relatives are found in Central
America and the Orient. In disturbed
areas and open woodlands, particularly those with pine, it is more often
heard than seen as it scurries off
in the leaf litter. True to its name,
it lives close to the surface, rarely
climbs, and needs little cover to hide
effectively. Most activity is during
warm, humid weather, but these
skinks may also be active on sunny
winter days.

A female, depending on her size,
lays from 1 to 7 eggs per clutch. An
egg is about 9 mm (0.4 in.) long, and
development is well under way when
laid. Two or more clutches may be
laid in a season, and early broods
mature within the year. Hatchlings
are about 45 mm (1.75 in.) long. This
is a short-lived species with numerous predators; about 10 percent of all
individuals survive as long as 2 years,
and few, if any, live a fourth year.
Small insects and other arthropods
are the primary foods.

Slender Glass Lizard
Ophisaurus attenuatus
559 to 1,180 mm (22 to 46.5 in.)
Glass Lizards are legless and closely resemble snakes but have movable eyelids, external ear openings, and a groove along each side of the body. The Slender Glass Lizard, our area's largest lizard species, has 2 to 4 dark lines below the lateral groove and often under the tail. The back is brown or tan, usually with a dark median stripe or a trace of one. Along each side are several black stripes and thin white or yellowish lines. Some large adults have pale dorsal crossbars with dark margins. The long, fragile tail is often broken, and the regenerated tip is light colored.

This species occurs in the southern half of Virginia east of the Blue Ridge Mountains, much of North Carolina, and most of South Carolina. It apparently is absent from most of the Mountains. Favorite habitats are grassy fields, woodland margins, and other open, usually dry places.

Slender Glass Lizards are extremely energetic. They thrash about when handled and are difficult to catch without breaking the tail. These reptiles eat mostly invertebrates—especially grasshoppers and other large arthropods—but they may also take small lizards and snakes and the eggs of ground-nesting birds. A female lays from 4 to 19 eggs per clutch and may attend them during incubation. Hatchlings are about 185 mm (7.25 in.) long.

Island Glass Lizard
Ophisaurus compressus
380 to 610 mm (15 to 24 in.) This glass lizard has a narrow head and 1 dark stripe on each side of the body through the third and fourth scale rows above the lateral groove. The back is usually unpatterned, but a median dark stripe is sometimes present; this may be broken into short segments. Small white spots,

generally 1 at the edge of each scale, mark the body anteriorly, and distinct white bars are usually present on the neck. There are no scales between the upper labials and the eye.

Island Glass Lizards prefer xeric habitats in coastal pine and maritime forests. They are found occasionally under tidal wrack on sandy beaches. This rare species is known in our area only from Charleston, Colleton, Georgetown, and Jasper counties, South Carolina.

The biology and ecology of this glass lizard are poorly known. Arthropods and other invertebrates probably constitute the primary prey. Hatchlings are smaller than those of other species. The caudal vertebrae lack fracture planes; hence the tail is less easily broken than in other glass lizard species.

Mimic Glass Lizard
Ophisaurus mimicus
380 to 660 mm (15 to 26 in.) This small glass lizard was long overlooked and only recently described. It closely resembles the Slender and Island Glass Lizards but differs in size and in a strong combination of scale characteristics, including a low number of scales along the lateral groove (usually 94 or fewer, compared with 100 or more in most other glass lizards). The tail (if unbroken) is very long, and the head is narrow. Dorsal coloration is tan to brown with a dark middorsal stripe that is most distinct posteriorly. Three or 4 dark stripes or rows of spots are present above the lateral groove. Markings below the groove are usually faint or absent. One or more upper labial scales usually reach the eye. Large males, especially, may have distinctive light and dark spots on the head, neck, and anterior body.

This apparently rare species is known in our area from southeast-ern North Carolina and the outer Coastal Plain of South Carolina. Virtually all specimens have been collected in relatively extensive tracts of flatwoods dominated by Long-leaf Pine and Wiregrass or on roads traversing such habitats.

Nothing is known of this species' reproductive biology in our area. Hatchlings have not been described. Arthropods and other invertebrates probably constitute the primary prey.

The Mimic Glass Lizard is listed as a species of Special Concern in North Carolina.

Eastern Glass Lizard (adult male)

Eastern Glass Lizard
Ophisaurus ventralis

457 to 1,067 mm (18 to 42 in.) This large legless lizard differs from the Slender Glass Lizard by lacking dark marks below the lateral groove and from the Island and Mimic Glass Lizards by having small scales between the eye and upper labials. It differs from all three by lacking a dark middorsal stripe. The dorsal color of young animals is usually brown or tan, and 1 or more dark stripes occur along each side of the body. Old adult males usually become greenish with heavy gold or bronze flecking. The belly is often bright yellow.

Probably the most frequently encountered glass lizard in our area, this species ranges throughout the Coastal Plain of South Carolina and in most of eastern North Carolina, northward to extreme southeastern Virginia. Favorite habitats are flatwoods, maritime forests, and other sandy environments. These lizards may be common under rubbish in fields and vacant lots near ponds, marshes, and estuaries.

Glass lizards are opportunistic feeders, taking mostly large arthropods and other invertebrates, but they also eat small vertebrates on occasion. Individuals forage chiefly

Eastern Glass Lizard (subadult)

by day, especially in the early morning and late afternoon, but are sometimes active at night. In late spring or summer, a female deposits from 5 to 15 eggs, usually in a shallow depression under a log or similar object, and she often remains with the eggs until they hatch. Hatchlings average about 170 mm (6.75 in.) in length.

The Eastern Glass Lizard is listed as a Threatened species in Virginia.

Suborder Serpentes

SNAKES

The earliest known snakes date from the late Cretaceous, about 80 million years ago. They unquestionably evolved from lizard ancestors, but fossil verification is still lacking. Unlike most lizards, snakes lack legs; however, some primitive families retain traces of the pelvic girdle, and a few have traces of legs. All lack movable eyelids, a middle ear, and osteoderms (bones in the skin). Snakes have extremely long vertebral columns, containing as many as 440 vertebrae. Adults range in size from 100 mm (4 in.) to about 10 m (33 ft.). Snakes cannot break off and regenerate the tail, as many lizards do. If the tail is broken, a snake remains stub-tailed for the remainder of its life.

All snakes are carnivores. Many swallow their prey alive, but some first wrap their body tightly around a prey item, causing suffocation and heart failure. Others have salivary glands specialized to produce venom and tubular teeth to inject it, thus immobilizing or killing the prey. Snakes do not chew their food; they swallow their prey whole. Their sharp, recurved teeth simply prevent prey from escaping. The jaw mechanism is unusually flexible. The quadrate bone is highly movable, and the right and left halves of the lower jaw are united by an elastic ligament, thus permitting large animals to be swallowed. Snakes contribute greatly to human well-being by controlling destructive pests, chiefly rats and mice.

This highly successful group comprises some 13 families, about 400 genera, and more than 2,700 species. In addition to burrowing in sand, soil, or mud, they are well adapted to life in freshwater or the open ocean, in trees, and on the ground. The majority of species inhabit the tropics, but many abound in the temperate zone. Most snakes (about 270 genera) are in the cosmopolitan family Colubridae, an enigmatic and complex assemblage.

Our area has a rich fauna containing 39 species, 33 of which are colubrids. *Nerodia* (water snakes), with 5 species, is the largest genus; *Lampropeltis* (kingsnakes), with 3, is next; 10 genera have 2 species each; and 11 genera are monotypic. Several species occur areawide, or nearly so, and none is endemic or even closely endemic. The Coastal Plain of the Carolinas has the most species (37), and the Mountains of Virginia have the fewest (22). Six of our species (3 rattlesnakes,

Copperhead, Cottonmouth, and Eastern Coral Snake) are venomous. Most species mate in April or May, but some mating occurs at other times of the year, particularly in the fall for many species. Most young appear in August or September. About half of our species lay shelled eggs, and the other half bear their young alive.

Eastern Worm Snake
Carphophis amoenus

191 to 330 mm (7.5 to 13 in.) Among the smallest of North American serpents, this species is characterized by its small, pointed head; plain chestnut to dark brown dorsum; and pinkish, translucent venter. Its tail tip bears a sharp but completely harmless spine. Dorsal scales are smooth and glossy. These snakes are appropriately named, for they indeed superficially resemble large worms. The largest individuals are usually females.

This species occurs in forested places, especially in moist environments, throughout Virginia and the Carolinas up to about 1,300 m (4,300 ft.) elevation.

Worm Snakes are highly secretive, spending most of the day beneath stones, logs, or other cover and prowling on the surface chiefly at night. They are excellent burrowers, and individuals are sometimes uncovered during excavation or gardening. Worm Snakes are inoffensive and do not bite. A restrained individual may pry against its captor's fingers with its head and its harmless tail spine. Earthworms constitute the principal food. Usually in June or July, a female deposits a clutch of from 2 to 8 eggs, frequently in a rotten log or mulch pile or beneath a large, flat stone. Hatching occurs in August or September. Hatchlings are about 100 mm (4 in.) long and are usually darker than adults, with brighter orange-red bellies.

Scarlet Snake *Cemophora coccinea*
356 to 660 mm (14 to 26 in.) These
small, smooth-scaled snakes are
brightly marked with red blotches
enclosed in black margins and
separated by white, yellow, or pale
gray interspaces. The undersurface
is white, translucent, and glossy. The
red snout projects well beyond the
lower jaw. This snake may superficially resemble the Eastern Coral
Snake, but the red snout, black
bands separating the red and yellow ones, and blotches on the back
rather than rings encircling the body
allow accurate identification. The
conical head is an adaptation for
burrowing.

Scarlet Snakes inhabit much of
the Coastal Plain and Piedmont but
are absent from most of the Mountains. Although their habitats are
varied, these snakes are most common in the sandhills and sandy pine
flatwoods of the Carolina Coastal
Plain.

Scarlet Snakes normally spend
daylight hours underground, prowling on the surface mostly on warm

nights, when they are often killed
on roads. These snakes virtually
never bite, even when first handled.
Reptile eggs constitute the bulk of
the diet. Small eggs are swallowed
whole, while the shells of larger eggs
are slit with the enlarged maxillary
teeth to access the contents. Lizards,
snakes, and other small animals
may occasionally be eaten; such
active prey may be restrained by constriction. The 2 to 6 eggs per clutch
are usually laid in July or August, and
hatchlings are about 130 mm (5 in.)
long. Much remains to be learned
about this secretive snake in our
area.

Black Racer (adult)

Black Racer *Coluber constrictor*
914 to 1,676 mm (36 to 66 in.) Adults
of this abundant and well-known
snake have a uniformly black dorsum and a predominantly black or
dark gray undersurface. The chin,
and sometimes the anterior venter,
is variously mottled with white.
Juveniles have conspicuous brown,
gray, or reddish dorsal blotches on a
paler ground color. The dorsal scales
are smooth.

Black Racers occur throughout the area up to about 1,400 m
(4,600 ft.) elevation. In the Mountains, they are most common in
valleys and at lower elevations. Habitats range from brushy dunes and
maritime forests to rocky hillsides
and upland meadows. Individuals
frequently hide beneath boards,
pieces of tin, or other surface cover
around rural buildings.

This strictly diurnal species is
among the fastest and most agile of
our snakes. When surprised, most
Racers crawl rapidly away, disappear-
ing down a nearby hole or into thick
vegetation. If cornered or restrained,
they vibrate the tail and actively defend themselves, sometimes making
wild lunges at the intruder. However,
their small teeth, like those of most
other nonvenomous snakes, inflict
only superficial cuts similar to those
resulting from briar scratches. Most
ordinary clothing provides more
than adequate protection against
the bites of these and other nonvenomous snakes in the region.
Insects, amphibians, reptiles, birds
and their eggs, and small mammals
are included in this snake's varied
diet. Despite its species name, it is
not a constrictor. Most Racers mate
in the spring, and in June or July a

Black Racer (juvenile)

female lays from 4 to 25 eggs that vary considerably in size and shape. Numerous small nodules on the shells give the appearance that the eggs have been sprinkled with salt. Like many snakes, Racers deposit their eggs under stones or in sawdust piles, rotten logs, stumps, and similar places. Several nests have been exposed by plows in sandy fields. Communal nesting occurs occasionally. Hatchlings emerge in late summer or early fall and are about 290 mm (11.5 in.) long.

Northern Ring-necked Snake

Ring-necked Snake
Diadophis punctatus

245 to 508 mm (10 to 20 in.) Easily identified by its bright yellow or orange collar, the Ring-necked Snake in our area is a small black or slate gray species with smooth scales and a yellow or orange undersurface. In most of South Carolina and the Coastal Plain of North Carolina and Virginia, Ring-necked Snakes have incomplete neck rings, a prominent row of black spots down the center of the belly, small black spots on the chin and some of the lower labials, 7 upper labials per side, and fewer than 190 ventrals plus subcaudals. Individuals from northern and western Virginia and from most of the Carolina Mountains usually have a complete neck ring, 8 upper labials per side, and more than 190 ventrals plus subcaudals. Belly, chin, and lower labial spotting is generally weak or absent in montane popu-

lations. The largest Ring-necked Snakes occur in the Mountains. Females often grow larger than males.

This species, common and widespread in most of the United States, ranges throughout the area below about 1,750 m (5,740 ft.) elevation. Ring-necked Snakes frequent forested habitats and are most often found beneath stones and in or under decaying logs and stumps, especially in moist places. Surface activity occurs chiefly at night, when individuals are sometimes found on roads.

When first captured, Ring-necked Snakes discharge potent musk and thrash about vigorously, but they rarely attempt to bite. Earthworms and small salamanders constitute the principal food, but these serpents also eat frogs, lizards, and small snakes. A female deposits from 2 to 11 eggs per clutch in old

Southern Ring-necked Snake

sawdust piles, rotten logs, or damp
soil under flat stones. As is the
case with many snakes, the largest
females often lay the most eggs.
Seven hatchlings from the Coastal
Plain of North Carolina averaged
108 mm (4.25 in.) in length, and 77
hatchlings from the North Carolina
Mountains averaged 130 mm (5 in.)
in length.

Corn Snake
Elaphe guttata (or Pantherophis guttatus)

762 to 1,829 mm (30 to 72 in.) This beautiful snake is glossy red, gray, orange, or brown, with prominent reddish brown blotches outlined in black. The venter is boldly checkered with black and white. There is a spear-shaped blotch on top of the head. The dorsal scales usually have weak keels. Males are often larger than females.

These snakes range over much of the area below about 760 m (2,500 ft.) elevation, but they are most common in pine and Wiregrass flatwoods and sandhills of the Carolina Coastal Plain.

Corn Snakes are secretive and frequently hide beneath surface cover, in stump holes, and in burrows of other animals. They may be active day or night but are most often nocturnal in hot weather. They climb with ease but are found most often on the ground. Many are unfortunately killed by cars. Recently captured specimens may bite, but most soon become docile. Their beautiful color patterns and tendency to thrive in captivity have made these snakes very popular in the pet trade.

Powerful constrictors, Corn Snakes feed mostly on small mammals. They also eat birds and their eggs, and frogs and lizards are favorite foods of juveniles. A female deposits a clutch of from 3 to 27 eggs, usually in June or July; these hatch in August or September. Hatchlings average about 320 mm (12.5 in.) in length and are usually less brightly colored than adults.

Black Rat Snake (adult)

Rat Snake
Elaphe obsoleta (or Pantherophis obsoletus)
1,067 to 2,159 mm (42 to 85 in.) This highly variable species is one of our largest and most abundant snakes. In most of the area, Rat Snakes have a black or dark brown dorsum and a venter mottled with gray and white. The anterior portion of the under-surface is lighter than the posterior portion and is often white. Individuals in much of the Coastal Plain of the Carolinas are greenish or yellow-ish with 4 prominent dark brown or black longitudinal stripes and gen-erally weak and diffuse ventral mark-ings. Juveniles have conspicuous gray or brown blotches on a lighter ground color, and some adults have traces of the blotched pattern. The dorsal scales are weakly keeled. Male Rat Snakes often attain larger sizes than females.

Rat Snakes range throughout Virginia and the Carolinas up to 1,350 m (4,430 ft.) elevation. Favorite habitats include upland hardwood forests, pocosins, river swamps and lowlands, fields, and barns and other buildings. Juveniles especially sometimes enter inhabited rural dwellings. These snakes are excellent climbers and frequently live in tree hollows, sometimes at considerable distances above the ground.

Yellow Rat Snake (adult)

Rat Snake (juvenile)

When approached, a Rat Snake often kinks its body and remains motionless. If provoked, it quickly assumes a defensive posture, vibrates its tail, and strikes. Small mammals and birds and their eggs are the principal foods of these large and powerful constrictors, but frogs and lizards are also eaten by young snakes. Because rats and mice constitute a major portion of the diet, Rat Snakes are economically beneficial to humans. However, on poultry farms and near nesting boxes for other birds, they may become a minor nuisance. The eggs, often laid in an adherent cluster, vary in number from 5 to 28 per clutch and are deposited in late spring or summer. Rotten logs and stumps, compost mounds, and sawdust piles are favorite nesting places. Hatchlings, averaging about 330 mm (13 in.) long, emerge in late summer or fall.

Mud Snake *Farancia abacura*
1,016 to 1,854 mm (40 to 73 in.) Mud Snakes are large and moderately stout-bodied with smooth, iridescent scales. Dorsal color is black; the undersurface, usually red, orange, or pink but occasionally white, has prominent black markings. Ventral ground color extends upward on the sides to form triangular bars. The lateral bars of some individuals, especially juveniles, extend as narrow crosslines across the back. The largest Mud Snakes are females.

Mud Snakes occur in the Coastal Plain and lower Piedmont of the Carolinas and in southeastern Virginia. They are mostly aquatic, inhabiting cypress swamps, sluggish lowland streams, ditches, and other heavily vegetated aquatic habitats.

When first caught, a Mud Snake virtually never bites but thrashes about and often presses its pointed but completely harmless tail tip against its captor's skin. This defensive behavior, also characteristic of the Rainbow Snake and Worm Snake, probably is the basis for the "sting-snake" myth of southern folklore. The "hoop snake" myth is also based on this species, but neither the Mud Snake nor any other snake is capable of rolling like a hoop. Amphiumas and sirens are the primary foods, but other amphibians may also be eaten. The eggs of this often prolific species number from 4 to 111 per clutch and are laid in June or July. Hatchlings appear in early fall, averaging about 220 mm (8.75 in.) long. Females usually attend their egg clutches during the incubation period.

Rainbow Snake
Farancia erytrogramma

914 to 1,676 mm (36 to 66 in.) This appropriately named species is one of our most colorful snakes. Dorsal ground color is blue-black with 3 longitudinal red stripes. The undersurface is mostly red or pink with 2 or 3 rows of black spots. Dorsal scales are smooth and iridescent. Females attain larger sizes than males.

Rainbow Snakes range over most

of the Coastal Plain, where habitats include rivers, large creeks, cypress swamps, lakes, canals, large ditches, and fresh and brackish marshes. In Virginia, especially during the spring, these animals have been exposed by plows in sandy fields near water. They may be active year-round.

Infrequently encountered, these snakes are highly aquatic, secretive, and chiefly nocturnal, hiding by day among cypress roots or under various debris. When first seized, a Rainbow Snake often presses its harmless tail tip against its captor's hand, but it seldom if ever bites. Amphibians are the main food of juveniles, and adults feed largely on American Eels (*Anguilla rostrata*). The eggs, often deposited in sand, number from 20 to 52 per clutch, and hatchlings are about 210 mm (8.25 in.) long. Females often remain with their eggs during incubation.

Eastern Hognose Snake (patterned adult with head and neck spread)

Eastern Hognose Snake
Heterodon platirhinos

508 to 1,194 mm (20 to 47 in.) This moderately large, stout-bodied snake has keeled dorsal scales and an up-turned snout with a pronounced median keel. Color patterns are highly variable. Usually the dorsum has dark blotches with light interspaces and varies from brown to shades of red, yellow, or orange, but some individuals are completely black or dark gray. This species differs from the Southern Hognose Snake by having the posterior part of the belly usually darker than the undersurface of the tail. It also attains a larger size, has a less upturned snout, and is proportionally more slender than the Southern Hognose. Females attain larger sizes than males.

Eastern Hognose Snakes range below about 760 m (2,500 ft.) elevation throughout most of the area, especially in habitats having sandy or friable loamy soils. This species is most common in the Coastal Plain and generally rare in the Mountains.

These reptiles are strictly diurnal and, unlike many snakes, are found crawling abroad more often than under sheltering objects. When alarmed, an individual usually flattens its head and neck and hisses loudly. If further provoked, it may strike (usually with its mouth tightly shut), void musk and feces, regurgitate food items, tightly coil its tail, gape its mouth, and eventually

Eastern Hognose Snake (profile of black phase specimen)

Eastern Hognose Snake (feigning death)

roll onto its back and feign death. Because of such extraordinary behavior, this snake is often rumored to be dangerously venomous, and such ominous names as "spreading adder," "black adder," "spreadhead moccasin," and "blowing viper" are firmly entrenched in the vocabulary of many rural persons. Surprisingly, a Hognose Snake almost never bites, even when first handled. Toads are the principal food, but other amphibians, as well as other small animals, are occasionally eaten. A female produces from 4 to 60 eggs per clutch, usually in June or July. Nests have been discovered a few centimeters below the surface in sandy fields. Hatchlings average about 180 mm (7 in.) in length.

Southern Hognose Snake
Heterodon simus

330 to 743 mm (13 to 29 in.) This small, exceptionally stout-bodied snake has keeled dorsal scales and a prominently upturned snout. Dorsal ground color is brown, tan, or gray, with dark blotches and pale interspaces. The belly is whitish, usually mottled with gray or brown. The undersurface of the tail is similar in color to the posterior part of the belly, unlike that of the Eastern Hognose Snake. Females attain larger sizes than males.

Southern Hognose Snakes inhabit much of the Coastal Plain of South Carolina and the Sandhills and southeastern Coastal Plain of North Carolina. Favorite habitats are sand ridges, pine and Wiregrass flatwoods, sandy old fields, and other xeric communities with coarse sands or porous, loamy soils.

When disturbed, these snakes may hiss, flatten the head and neck, and feign death, but they usually do so with less enthusiasm than Eastern Hognose Snakes. They virtually never bite. Southern Hognose Snakes are effective burrowers, spending much time underground. Surface activity occurs almost exclusively during the day, and the species is most active in the fall. Adults in the North Caro-

lina Sandhills have relatively small home ranges, exhibiting little site fidelity and overwintering in burrows usually of their own making, at depths of about 25 to 46 cm (10 to 18 in.). Toads, especially spadefoots, are the primary food, but other amphibians and lizards are also taken. Mating occurs in both spring and fall. Females deposit from 6 to 15 eggs, usually in July. Natural nests have not been reported. Hatchlings appear in September and October, averaging about 155 mm (6 in.) in length.

The Southern Hognose Snake is declining throughout its range due to habitat loss and fragmentation and possibly predation by the Red Imported Fire Ant. It is protected as a species of Special Concern in North Carolina and as a species In Need of Management in South Carolina.

Mole Kingsnake
Lampropeltis calligaster

762 to 1,168 mm (30 to 46 in.) This moderately large snake has a series of small, dark blotches on a brown, tan, or reddish dorsum. Juveniles and young adults have bright markings; but some old individuals have very faint ones, and old males in particular often have unpatterned brown backs and yellowish bellies. The dorsal scales are smooth, and the head is comparatively small. Males often reach larger sizes than females.

These kingsnakes occur over much of the area, but their distribution in some sections is poorly known. They are apparently absent from most of the Mountains and from extreme southeastern Virginia and northeastern North Carolina, and only a few records are known from the lower Coastal Plain of South Carolina. Favorite habitats range from sandhills and pine flatwoods to open fields and upland hardwood forests. Although these snakes are accomplished burrowers, they spend much time prowling aboveground and under logs and other surface cover. They may be active day or night and are frequently found on roads.

When disturbed, a Mole Kingsnake vibrates its tail and may attempt to bite. Small mammals, lizards, and snakes are the principal food of these constrictors. A female deposits from 5 to 17 eggs, usually in June or July. Nests have been found a few centimeters below the surface in sandy fields. Hatchlings average about 225 mm (8.75 in.) long.

Eastern Kingsnake
Lampropeltis getula
914 to 1,753 mm (36 to 69 in.)
This large, handsome serpent has smooth, glossy scales. Most specimens are black with bright white or yellow chainlike markings. Along the Outer Banks of North Carolina, from Cape Hatteras to Cape Lookout, these snakes usually are brownish with conspicuous light stippling in the dark areas between the chainlike patterns. In extreme southwestern Virginia, adults are mostly black with only a faint trace of the chainlike markings, while juveniles have the chain pattern. The largest Eastern Kingsnakes are usually males.

Eastern Kingsnakes occur below about 760 m (2,500 ft.) elevation in most of the area, but they are most common in the Coastal Plain. Preferred habitats include banks of rivers and canals, river floodplains, and edges of old fields and forests. Though active on the surface day and night, they are frequently secretive, often hiding under surface cover, especially around trash dumps and old buildings.

When first caught, a Kingsnake usually expels copious quantities of musk from its cloacal glands and may bite, but most soon become docile and tolerate handling. These powerful constrictors eat turtle eggs, lizards, birds, small mammals, and other snakes, including venomous ones. The eggs, 5 to 20 per clutch and often adherent, are laid in decomposing logs and similar places in June and July. Hatchlings emerge from late August to early October and are about 270 mm (10.5 in.) long.

Eastern Kingsnakes are apparently declining over much of their range. The Outer Banks population is protected in North Carolina as a taxon of Special Concern.

Eastern Milk Snake

Eastern Milk Snake, Scarlet Kingsnake
Lampropeltis triangulum

This wide-ranging, highly variable species is represented in Virginia and the Carolinas by two subspecies so different that for years they were considered separate species (some authorities still advocate recognizing them as such). A zone of intergradation in which snakes have characteristics intermediate between the Scarlet Kingsnake and the Eastern Milk Snake extends from northeastern North Carolina through an arc in Virginia from the south central counties through the Coastal Plain to the Eastern Shore, but the two may occur sympatrically without interbreeding in some portions of the North and South Carolina Mountains.

610 to 1,143 mm (24 to 45 in.) The Eastern Milk Snake (*L. t. triangulum*) has large red, brown, or gray body blotches; a pale Y-, U-, or V-shaped marking on the nape; and a black and white checkered venter. Juveniles typically have redder blotches than adults. Milk Snakes occur in western Virginia and the Mountains of the Carolinas up to about 1,500 m (4,920 ft.) elevation. They prefer woodlands but also inhabit grassy balds and meadows. Much time is spent under cover, but individuals may be active on the surface both day and night. Milk Snakes sometimes enter rural buildings, even those inhabited by humans. From 5 to 20 often adherent eggs are laid in decaying logs or in loose soil under objects; hatchlings average about 210 mm (8.25 in.) long.

Scarlet Kingsnake

356 to 686 mm (14 to 27 in.) The beautiful Scarlet Kingsnake (*L. t. elapsoides*) has a bright pattern of red, black, and yellow or white bands that usually encircle the body. It differs from the highly venomous and superficially similar Eastern Coral Snake by having a red snout and contiguous red and black bands, as opposed to a black snout and contiguous red and yellow bands. Scarlet Kingsnakes range from southeastern Virginia through most of the Carolinas; but records are scarce in much of the Piedmont, and only a few specimens are known from the Mountains. These snakes are most common in pine flatwoods in the Carolina Coastal Plain, where they often hide under loose bark and inside logs and stumps. More secretive than Eastern Milk Snakes, they are active on the surface primarily at night. Scarlet Kingsnakes lay from 3 to 7 eggs, usually in rotten wood. Hatchlings are about 150 mm (6 in.) long and emerge in late summer.

Scarlet Kingsnakes and Eastern Milk Snakes are smooth-scaled constrictors that feed chiefly on lizards, small snakes, and small mammals. Scarlet Kingsnakes rely heavily on skinks as prey.

Eastern Coachwhip (adult)

Eastern Coachwhip
Masticophis flagellum (or Coluber flagellum)

1,219 to 2,388 mm (48 to 94 in.)
The longest snake in our area, the slender-bodied Coachwhip is dark brown to black on about the anterior third of its body, grading to pale brown or tan on the remainder. The scale pattern on the posterior body and tail appears braided, like a whip. The dorsal scales are smooth. Juveniles differ from adults by having dorsal patterns of wavy dark crosslines that are most conspicuous anteriorly, 2 longitudinal rows of brown to black spots on the anterior venter, and white outlines around the large scales on top of the head. Each parietal scale has a large pale spot, and the sides of the head are prominently mottled with white. Male Coachwhips usually reach larger sizes than females.

This species ranges through most of South Carolina and the Sandhills and southeastern Coastal Plain of North Carolina. Favorite habitats are sandy flatwoods, maritime forests, and sandhills with pines, scrub oaks, and Wiregrass. These snakes often occur on grassy dunes very near the ocean.

The Coachwhip's habit of frequently prowling with its head and neck raised well above the ground is unusual among local snakes. Diurnal and extremely active, it is perhaps the fastest and most agile of our serpents. When encountered in the open, a Coachwhip crawls rapidly for the nearest hole or thick vegetation, or it may climb a shrub or small tree. However, it rapidly vibrates its tail and bites vigorously and repeatedly if cornered or restrained. Some individuals may exhibit thanatosis, or death-feigning.

Eastern Coachwhip (juvenile)

Despite rural folklore, Coachwhips do not whip or thrash victims with their tails. Lizards, snakes, birds and their eggs, and small mammals are the principal foods. Racerunners are regularly eaten, although these swift lizards probably are caught primarily by ambush or when inactive rather than by pursuit. Female Coachwhips deposit from 4 to 15 eggs per clutch in late spring or summer. Their shells, like those of the Black Racer, bear numerous small granules resembling grains of salt. Hatchlings are about 420 mm (16.5 in.) long. Coachwhips in the North Carolina Sandhills have exhibited fidelity toward hibernacula and other refugia within well-defined home ranges.

The Coachwhip has suffered declines over much of its range due to habitat loss and fragmentation and possibly to nest predation by the Red Imported Fire Ant. Much remains to be learned about this snake's natural history in our area.

Red-bellied Water Snake (adult)

Red-bellied Water Snake
Nerodia erythrogaster

762 to 1,524 mm (30 to 60 in.) Adults of this large, keel-scaled water snake usually have a uniformly reddish brown to dark brown dorsum, a plain orange to reddish belly, and a white patch on the chin. Occasional specimens have white mottling and scattered dark patches on the belly. Females usually reach larger sizes than males. Juveniles closely resemble Northern Water Snakes by having dark brown crossbands on the neck and anterior body and dorsal blotches with alternating lateral bars on the posterior body. The dorsal ground color is pinkish, and the venter is pale orange to pinkish, often with short brown or black bars along the edges. In contrast, Northern Water Snakes have prominent

ventral patterns of brown or reddish half-moon-shaped spots, frequently with narrow dark brown to black margins.

The Red-bellied Water Snake occurs in the Coastal Plain and along the Fall Line from southeastern Virginia throughout the Carolinas. A fingerlike projection of the range into northwestern South Carolina to Greenville County may well have resulted from the formation of suitable habitats by recent impoundments along rivers. Favorite habitats include river swamps, marshes, lakes, and other bodies of usually still or sluggish water, but these snakes sometimes move considerable distances from aquatic environments, especially during humid periods. Individuals away from water are most often found under various

Red-bellied Water Snake (juvenile)

kinds of surface cover around trash
dumps and abandoned buildings.

Like most members of the genus
Nerodia, Red-bellied Water Snakes
vigorously defend themselves by bit-
ing and discharging a foul-smelling
musk. Fish, toads, and frogs constitute the principal foods. Mating generally takes place in the spring. From
5 to 55 young per litter, averaging
about 255 mm (10 in.) in total length,
are born in August or September.

Banded Water Snake
Nerodia fasciata

610 to 1,397 mm (24 to 55 in.) This extremely variable species differs from the closely related Northern Water Snake by having black, brown, or reddish crossbands, usually unbroken throughout the length of the body; a dark stripe from the eye to the angle of the jaw; and squarish or triangular dark markings on the venter. Juveniles have bold patterns, whereas large adults often have obscure ones. The dorsal scales are strongly keeled. Females usually grow larger than males.

This snake is common in most permanent and semipermanent freshwater environments in the Coastal Plain of the Carolinas south of Albemarle Sound. It is known to hybridize with the Northern Water Snake in areas where the ranges of the two overlap. Banded Water Snakes are not known from Virginia, but some Northern Water Snakes from the southeastern corner of that state strongly resemble this species.

The habits of this snake are similar to those of many other water snakes. Individuals usually flee at the slightest disturbance, but if restrained or cornered, they bite vigorously and void musk from the cloacal glands. However, their bites, like those of other local nonvenomous snakes, produce only superficial scratches. Banded Water Snakes feed chiefly on fish and amphibians. Females give birth to from 6 to 82 young in late summer or early fall. Newborns are about 220 mm (8.75 in.) long.

Florida Green Water Snake
Nerodia floridana

762 to 1,880 mm (30 to 74 in.) The largest of our water snakes, this heavy-bodied species is dark green, greenish brown, olive, or brown with keeled scales. Some individuals have faint dark crossbars, but most lack distinct markings. The belly is uniformly whitish or cream colored; under the tail are pale spots on a gray or brown background. The presence of one or more small scales between the eye and the upper labials distinguishes this species from other water snakes in the area. Females attain larger sizes than males.

Green Water Snakes prefer quiet waters of streams, lakes, ponds, Carolina bays, and marshes. Most individuals have been captured in abandoned rice fields and rice field reservoirs. This species occurs in our area only in southern South Carolina, from Aiken and Barnwell counties southeastward. It is primarily active during warm months but may bask on warm days in winter.

When caught, Florida Green Water Snakes void potent musk and often bite vigorously. Adults are capable of inflicting painful but superficial scratches. Fish and frogs constitute the principal food. This species is among the most reproductively prolific of our snakes. The 7 to 132 young per litter are born in late summer and average about 240 mm (9.5 in.) long.

Northern Water Snake

Northern Water Snake
Nerodia sipedon

610 to 1,270 mm (24 to 50 in.) This extremely variable species may be brown, tan, gray, or reddish, usually with dark crossbands on the neck and anterior body and dark dorsal blotches and alternating lateral bars on the remainder of the body. The belly is yellowish, usually with brown or reddish half-moon-shaped spots frequently outlined with dark brown or black. Dorsal patterns are especially prominent in juveniles but often obscure in large adults. Females usually reach larger sizes than males. The dorsal scales are strongly keeled. In Virginia and most of North Carolina, these snakes usually have dark markings that are as wide as or wider than the spaces between them. Snakes in populations along the Potomac River may be uniformly black. Specimens from South Carolina generally have dark markings that are smaller than the interspaces, especially along the sides. Individuals living in and near brackish water along the Outer Banks and adjacent mainland of North Carolina are very dark dorsally, sometimes almost black, with black half-moons over most of the venter. This population is regarded as a distinct subspecies (*N. s. williamengelsi*, the Carolina Salt Marsh Snake or Carolina Water Snake) and is listed as a taxon of Special Concern in North Carolina.

Habitats range from mountain lakes and brooks below about 1,500 m (4,920 ft.) elevation to large

Carolina Water Snake

coastal estuaries, but this species is absent from most of southeastern North Carolina and the Coastal Plain of South Carolina. It is known to hybridize with the Banded Water Snake in areas where the ranges of the two overlap.

These abundant snakes often bask on logs and other debris in the water and along its edge. They frequently also climb among low, overhanging limbs. Like other water snakes, these serpents have short tempers and are often wrongly believed to be venomous. Most individuals bite readily and void strong musk when cornered or seized. Fish and amphibians are the chief foods.

Contrary to popular belief, water snakes are not detrimental to fish populations. Instead, they probably contribute to better fishing by feeding largely on stunted or diseased fish. Moreover, young water snakes provide excellent food for larger game fish. The 8 to 50 young per litter, born usually in later summer, are about 215 mm (8.5 in.) long at birth.

Brown Water Snake
Nerodia taxispilota

813 to 1,626 mm (32 to 64 in.) These large, heavy-bodied snakes have wide, flat heads, protruding eyes, and strongly keeled dorsal scales. The dorsum is brown with squarish black blotches, one series along the middle of the back and another alternating series along each side. The yellowish venter is prominently mottled with brown or black. Females attain larger sizes than males.

Brown Water Snakes range from eastern Virginia and North Carolina through much of South Carolina. They are most abundant in the lakes and larger streams of the Coastal Plain.

This species is the most arboreal and the most aquatic of our water snakes. On sunny spring days, numerous individuals bask on limbs over the water, into which they drop when disturbed. They are typically less wary than other water snakes, often permitting close approach and occasionally dropping into passing boats. Unfortunately, many are killed by fishermen who believe them to be Cottonmouths. Like most members of the genus, Brown Water Snakes usually bite and discharge musk when handled or restrained. Fish, especially catfish, are the primary food. This snake mates usually in the spring, often among branches overhanging the water. From 9 to 61 young per litter are born in summer or early fall. Neonates are about 290 mm (11.5 in.) long.

Rough Green Snake
Opheodrys aestivus

559 to 914 mm (22 to 36 in.) These slender, long-tailed snakes have a uniformly green dorsum and a plain yellow or greenish yellow belly. Hatchlings and small juveniles are gray or grayish green with whitish undersurfaces. The dorsal scales are keeled. The largest individuals are usually females.

Rough Green Snakes inhabit forested environments below about 1,100 m (3,600 ft.) throughout most of the area. They are agile climbers but are also often seen on the ground, and many are killed on roads. A favorite habitat is among the foliage of shrubs, vines, and small trees overhanging streams and lakes, especially in the Coastal Plain.

A Rough Green Snake, when first caught, may gape its mouth, exposing the black interior, but it seldom attempts to bite. Spiders, grasshoppers, crickets, caterpillars, and other soft-bodied arthropods are the principal foods; small snails are also eaten. A female lays from 3 to 12 elongated, often adherent eggs per clutch, usually in June or July. Communal nesting sometimes occurs; one such nest in North Carolina contained 74 eggs. Hatchlings averaging about 200 mm (7.75 in.) long emerge in August or September.

Smooth Green Snake
Opheodrys vernalis (or *Liochlorophis vernalis*)

356 to 610 mm (14 to 24 in.) The bright green dorsum and yellowish white venter make identification easy. This species differs from the Rough Green Snake by having smooth dorsal scales, by being smaller and stockier, and by having a more northerly range. After death, the green pigment of both species quickly fades to a pale blue. The young are grayish green above and whitish below.

Smooth Green Snakes inhabit moist, grassy meadows and fields in

northwestern Virginia. A specimen reportedly taken near the French Broad River in Madison County, North Carolina, in 1871 has been correctly identified, but the locality has been questioned. There are several other reports from that state, but none are supported by specimens.

Smooth Green Snakes may climb in shrubs and bushes but are most often found under stones and logs in high-elevation grassy habitats. They eat insects, slugs, millipedes, centipedes, spiders, and small salamanders. Smooth Green Snakes seldom bite, even when first caught. The eggs, laid usually in June and July, vary in number from 2 to 12 per clutch. Each egg resembles a blunt-ended cylinder, and its shell is thin and parchmentlike. Hatchlings emerge by early fall and are about 130 mm (5 in.) long.

Although of uncertain occurrence in North Carolina, the Smooth Green Snake is listed as a species of Special Concern there.

Pine Snake *Pituophis melanoleucus*
1,067 to 1,829 mm (42 to 72 in.) This
large, moderately stout snake has
4 prefrontal scales and keeled dorsal
scales. Most specimens are whitish
with dark brown or black blotches
that are most distinct on the posterior part of the body. In southern
South Carolina, Pine Snakes frequently are tan or rusty brown, often
with indistinct body blotches. The
largest individuals are usually males.

Pine Snakes occur in disjunct
populations, and much remains to
be learned about their distribution
in the area. Sandhills with pines and
scrub oaks, and sandy flatwoods
are the preferred habitats in the
Carolinas. These snakes inhabit dry
upland forests and shale barrens in
western Virginia, but almost nothing
is known about them in that region.

These handsome serpents spend
much time underground, with
stump cavities and mammal burrows serving as important refugia.
Individuals in the North Carolina
Sandhills often exhibit fidelity for
hibernacula and other refugia within
large but well-defined home ranges.
Most surface activity occurs during
the day. If cornered or restrained, a
pine snake typically hisses loudly,
vibrates its tail, and may bite. This
powerful constrictor feeds largely on
small mammals; but birds and their
eggs are also eaten, and lizards may
be taken by juveniles. A Pine Snake's
large eggs vary in number from 3 to
12 per clutch and are laid usually
in sandy, open areas. In the North
Carolina Sandhills, females excavate
their own nest burrows, a task that
may take hours or days. Nesting
behavior is poorly known in other
portions of our region. Hatchlings
are about 440 mm (17.25 in.) long.

Pine Snakes have suffered from
habitat loss and fragmentation. They
are protected as a species of Special
Concern in North Carolina.

Glossy Crayfish Snake
Regina rigida

356 to 775 mm (14 to 30.5 in.) These small to medium-sized snakes have a shiny brown or dark olive dorsum with faint darker stripes, dark cream or orange-brown lips, and a yellowish undersurface with 2 rows of prominent crescent-shaped black spots. The dorsal scales are keeled. Females typically grow larger than males.

This generally uncommon species ranges in the Coastal Plain

of the Carolinas south of Albemarle Sound and as a disjunct population in New Kent County, Virginia. Glossy Crayfish Snakes are largely aquatic and live in creeks, freshwater marshes, cypress ponds, Carolina bays, and sphagnum-choked canals, but individuals are seen most often at night on roads through or near these habitats.

When handled, these snakes flatten their heads and bodies and void an especially pungent musk from the cloacal glands, and some may attempt to bite. Crayfish apparently constitute the bulk of the diet, but dragonfly naiads and other aquatic insects, as well as occasional small vertebrates, may also be eaten. Females bear litters of from 6 to 14 young in late summer. Neonates are about 190 mm (7.5 in.) long. Little is known of the ecology of this seldom-encountered species in our area.

Queen Snake *Regina septemvittata*
381 to 864 mm (15 to 34 in.) This
slender, highly aquatic snake has
keeled scales, a small head, and yellow lips. Its dorsum is dark brown,
sometimes with 3 usually obscure
black stripes. A yellow stripe along
each lower side involves the first and
second scale rows. The yellowish
venter, variously striped or mottled
with brown or tan, has a dark brown
stripe along each side. Juveniles
generally have brighter patterns than
adults, and females grow larger than
males.

In our area, this species occurs
primarily in rocky streams and rivers
in the Piedmont and Mountains up
to about 760 m (2,500 ft.) elevation.
It is also known from some portions
of the Coastal Plain and Sandhills.

Largely diurnal, these snakes
frequently bask on limbs over the
water, but they are found most often
beneath stones and debris along
the water's edge. When first caught,
a Queen Snake usually thrashes
about and discharges musk. Some
individuals may bite. These snakes
specialize in feeding on freshly
molted ("soft") crayfish. The number
of young per litter varies from 5 to
15, and neonates are about 200 mm
(7.75 in.) long.

Pine Woods Snake
Rhadinaea flavilata

254 to 381 mm (10 to 15 in.) These small snakes have a golden brown or reddish brown dorsum and a glossy yellowish or whitish venter. The top of the head is darker than the general ground color, and a brown line extends on each side of the head posterior to the eye. The dorsal scales are smooth and iridescent.

This highly secretive, infrequently encountered species primarily inhabits the lower Coastal Plain of the area north to Bodie Island on the Outer Banks in Dare County, North Carolina. It also ranges as far inland as Aiken County, South Carolina, and Scotland County, North Carolina. The common name is appropriate, for these serpents often occur under and inside rotten logs and stumps in pine flatwoods and in ecotones between sandhills and bottomlands. They are active on the surface chiefly at night.

The Pine Woods Snake has enlarged rear teeth in the upper jaw and apparently secretes a weak venom that immobilizes the small frogs and lizards on which it feeds. These snakes are, however, entirely harmless to humans and do not bite, even when first caught. They do, however, exude a particularly potent, amber-colored musk from their cloacal glands. The eggs are capsule-shaped, number from 2 to 4 per clutch, and usually are laid in decaying wood from June to August. Hatching occurs from August to October, and hatchlings are about 140 mm (5.5 in.) long.

Black Swamp Snake
Seminatrix pygaea

254 to 483 mm (10 to 19 in.) Small size, glossy black dorsum, and red or orange undersurface with conspicuous black bars along the scale edges characterize this species. Dorsal scales are smooth, although they sometimes have thin, light lines that superficially resemble keels. Females attain larger sizes than males.

This species occurs in the Coastal Plain of South Carolina and the lower Coastal Plain of North Carolina south of Albemarle Sound, and on Bodie Island on the Outer Banks. Favorite habitats are cypress ponds, sphagnum-choked canals, ditches, and sluggish lowland streams with lush vegetation.

Much remains to be learned about this secretive and essentially aquatic snake in our area. Although apparently uncommon to rare over most of our area, it has been found in abundance in some Carolina bay habitats in the South Carolina Coastal Plain. Individuals have been collected among aquatic plants, in sphagnum moss, and beneath various debris along the water's edge. They sometimes occur on roads at night, especially after heavy summer rains. Black Swamp Snakes usually do not bite, even when first handled. Worms, leeches, small fish, and amphibians are the chief foods. Born in late summer, the 2 to 22 young per litter are about 150 mm (6 in.) long.

Brown Snake *Storeria dekayi*
229 to 457 mm (9 to 18 in.) The dorsum of this small serpent is brown, gray, or reddish, usually with a pale median stripe. Small, paired black or dark brown spots usually occur along the margins of the stripe. The spots of some individuals fuse to form short transverse bars. The belly is white to pinkish, sometimes with tiny black spots along each side. A newborn juvenile has a dark,

virtually unpatterned dorsum and a prominent whitish neck band. Dorsal scales are keeled. Females attain larger sizes than males.

This species, widely distributed in the eastern half of the United States, ranges through most of Virginia and the Carolinas up to about 1,220 m (4,000 ft.) elevation. Habitats vary from coastal flatwoods and drier pocosins to upland hardwood forests, but these snakes are most often found under paper, boards, or other debris in urban and suburban areas.

When handled, Brown Snakes usually do not bite but flatten their heads and bodies and expel musk from their cloacal glands. Slugs and earthworms are the chief foods. The 3 to 26 young per litter are born usually from late July to early September and are about 100 mm (4 in.) long at birth.

Red-bellied Snake
Storeria occipitomaculata

165 to 305 mm (6.5 to 12 in.) This diminutive snake has keeled scales, a small but usually conspicuous white blotch on the next-to-last upper labial, and 3 orange or yellowish nape spots, most often separate but sometimes fused. Frequently there is a light middorsal band bordered on each side by a black line, and some of the dorsal and lateral scales may have tiny white flecks. Most of the undersurface is red or orange, often with black or dark gray stippling forming a stripe along each side. The chin is whitish, heavily stippled with gray or black. Red-bellied Snakes are highly variable. Some have a tan or brown dorsum, others are gray or almost black, and some individuals from the Coastal Plain of the Carolinas are glossy reddish or reddish

orange dorsally and ventrally. Juveniles are usually darker and have brighter nape spots than adults, and females typically grow larger than males.

Widely distributed in the eastern United States, Red-bellied Snakes occur below about 1,700 m (5,580 ft.) elevation through most of Virginia and the Carolinas. They apparently are most abundant in upland sec-

tions, but habitats include pine and Wiregrass flatwoods, sandhills, swamp margins, open deciduous forests, timbered hillsides, and wooded residential areas. Specimens are found most frequently under stones, logs, and similar sheltering objects. They are largely nocturnal but may also be active by day.

A Red-bellied Snake when handled flattens its head and body, curls its upper lips in a sneerlike manner, and discharges musk from its cloacal glands; but it does not bite. Small slugs and snails constitute the principal food. Matings of this live-bearing species have been reported in spring and fall. The 2 to 16 young per litter, born usually in summer or early fall, are about 77 mm (3 in.) long at birth.

Southeastern Crowned Snake
Tantilla coronata

203 to 305 mm (8 to 12 in.) This diminutive snake has a black or dark brown head and collar separated by a light band across the rear of the head. The rest of the dorsum is uniformly tan or reddish brown, and the translucent belly is plain white or yellow. The dorsal scales are smooth. The largest individuals are usually females.

This species occurs throughout South Carolina and in most of North Carolina below 610 m (2,000 ft.) elevation. It is known in Virginia only from a few counties in the southern Piedmont and Coastal Plain. Preferred habitats are pine flatwoods, maritime forests, sandhills, and wooded slopes.

These secretive snakes normally spend the day underground or beneath surface debris and prowl on the surface at night. They feed primarily on centipedes, but some beetle larvae and other soft-bodied arthropods may also be eaten. Prey is subdued with a weak venom delivered via an enlarged and grooved rear tooth on each side of the upper jaw, but Crowned Snakes are completely harmless to humans and do not bite when handled. A female deposits 1 to 4 elongated eggs per clutch, usually in June or July. Hatchlings emerge in September or October and average about 105 mm (4 in.) long.

Eastern Ribbon Snake
Thamnophis sauritus

457 to 965 mm (18 to 38 in.) These slender, long-tailed snakes usually have 3 conspicuous yellow stripes on a dark brown dorsum and a white or yellow spot in front of the eye. Along each side, one of the stripes occupies the third and fourth scale rows. Dorsal scales are keeled. Most individuals have 7 upper labials per side and a prominent middorsal stripe. In extreme southeastern South Carolina, however, many have 8 upper labials per side and a less prominent or incomplete middorsal stripe. The largest Ribbon Snakes are usually females.

These semiaquatic snakes occur through most of the area below about 760 m (2,500 ft.) elevation. They are fairly common in the Coastal Plain and usually uncommon to rare in the Mountains and Piedmont. Favorite habitats are marshes, damp meadows, and stream margins.

Ribbon Snakes are active and extremely agile. When alarmed, they usually disappear quickly amid vegetation. When seized, they thrash about and void musk from their cloacal glands. Some individuals may bite. Amphibians and small fish are the principal foods. Mating has been observed in the fall and probably occurs in spring also. Three to 16 young are born in August or September, and juveniles are about 205 mm (8 in.) long at birth.

Eastern Garter Snake
Thamnophis sirtalis

457 to 1,067 m (18 to 42 in.) The Garter Snake is a highly variable, moderately large species with strongly keeled dorsal scales. Dorsal ground color may be various shades of green, blue, brown, or red. Often there is a conspicuous pale mid-dorsal stripe and a less prominent light stripe involving the second and third scale rows along each side, but some individuals lack stripes and have spotted patterns. Females attain larger sizes than males.

These familiar snakes occur throughout the region, but they are most abundant in the Mountains, where individuals range to or near the summits of the highest peaks. Their habitats are varied; but they are usually associated with moist environments, and specimens are often found under stones and other surface cover.

A Garter Snake, when first caught, flattens its head and body, often bites, and expels a pungent musk from its cloacal glands. Earthworms, fish, and amphibians form the bulk of the diet. The 5 to 101 young per litter are born in late June through September; neonates average about 175 mm (7 in.) long.

Rough Earth Snake
Virginia striatula

178 to 318 mm (7 to 12.5 in.) These small, secretive snakes have pointed snouts and strongly keeled dorsal scales in 17 rows around the body. The dorsum is plain brown or grayish brown, often with a pale band across the top of the head. The undersurface is glossy white or greenish white. Juveniles are usually

darker and have a more prominent head band than adults. The largest individuals are usually females.

The Rough Earth Snake occurs throughout the Coastal Plain and in much of the Piedmont of the Carolinas and in the lower half of the Coastal Plain in Virginia. Preferred habitats range from pine and Wiregrass flatwoods to suburban gardens and urban lots. Sometimes aggregations of these snakes can be found under stones, logs, and similar cover, especially in the spring.

When handled, a Rough Earth Snake extrudes musk from its cloacal glands, but it does not bite. Earthworms are the principal food. Mating in this viviparous species usually occurs in the spring, and from 2 to 13 young are born in July or August. The young at birth are about 95 mm (3.75 in.) long.

Smooth Earth Snake
Virginia valeriae

178 to 330 mm (7 to 13 in.) This small, moderately stout-bodied snake has a gray, brown, or reddish brown dorsum that may have small, scattered black spots. The belly is whitish and unmarked. The Smooth Earth Snakes from Highland County, Virginia, however, represent the southernmost population of a race (the Mountain Earth Snake, *V. v. pulchra*) with weakly keeled scales in 17 rows at midbody (other populations have smooth scales in 15 rows). The largest Smooth Earth Snakes are usually females.

This species probably occurs in most of the Carolinas and Virginia below about 610 m (2,000 ft.) elevation, but it is generally uncommon. Records are spotty in many places and absent from western Virginia and northwestern North Carolina.

These snakes are most often found on roads at night or beneath logs, stones, and similar surface cover in open woodlands, along forest edges, and in wooded residential areas.

This inoffensive little snake feeds chiefly on earthworms. When handled, it voids musk from its cloacal glands but does not bite. The 2 to 12 young per litter are born from late July to mid-September and are about 90 mm (3.5 in.) long at birth.

Eastern Coral Snake
Micrurus fulvius

508 to 890 mm (20 to 35 in.) This beautiful but highly venomous smooth-scaled snake has a black snout and colorful patterns of red, yellow, and black rings around the body. Each red and black ring is separated by a yellow ring, and the tail is banded with black and yellow.

Coral Snakes occur in the Coastal Plain of South Carolina and in southeastern North Carolina. Sandy flatwoods, maritime forests, and sandhills with pines, scrub oaks, and Wiregrass are typical habitats.

These secretive snakes burrow in sand and ground litter and prowl on the surface chiefly in the morning and late afternoon. Their venom is extremely toxic and may produce paralysis and respiratory failure, but individuals are not aggressive and usually bite only if handled or otherwise restrained; bites are virtually unknown in our region. Small snakes are the principal prey; some lizards may also be eaten. The number of eggs per clutch varies from 2 to 9, and hatchlings are about 200 mm (8 in.) long.

Much remains to be learned about this rare snake in our area. It is protected as an Endangered species in North Carolina.

Copperhead (adult)

Copperhead *Agkistrodon contortrix*
610 to 1,220 mm (24 to 48 in.)
Copperheads are moderately large,
stout-bodied, venomous snakes with
a conspicuous pit on each side of
the head between the eye and nostril. They have elliptical pupils that
in dim light become almost round.
Most subcaudal scales are undivided. Dark dumbbell- or hourglass-
shaped crossbands with light centers
constitute the body pattern, and the
dorsal scales are keeled. Juveniles
have greenish yellow to bright yellow
tail tips. Some Copperheads in the
Coastal Plain of the Carolinas and
southeastern Virginia have a pale,
often pink or tan ground color and
middorsal crossbands that are very
narrow (2 or 3 scale-lengths wide). In
most of Virginia, and in the Mountains and upper Piedmont of the

Carolinas, these snakes frequently
have a brown or grayish brown dorsal ground color, and the middorsal
crossbands are less constricted
(more than 3 scale-lengths wide).
The largest Copperheads are usually
males.

In most of the eastern United
States, the Copperhead is the most

Copperhead (juvenile showing yellow tail)

common venomous snake. With the exception of the highest mountain peaks, the most heavily urbanized areas, and the Outer Banks north of Bogue Banks, it ranges throughout Virginia and the Carolinas in a variety of habitats, from coastal flatwoods and drier pocosins to wooded slopes up to 1,400 m (4,600 ft.) elevation. In some parts of the region, it is the only venomous serpent. Individuals often hide beneath boards, pieces of tin, and similar objects around dilapidated rural buildings and trash dumps.

Most venomous snakebites in our area are from Copperheads, but these snakes are seldom aggressive and usually bite only if stepped on or otherwise provoked. Their bites, although rarely if ever fatal to humans, are serious and should receive prompt medical attention. Copperheads eat large insects, amphibians, reptiles, birds, and small mammals. They may be active both day and night but are mostly nocturnal during hot weather. Mating occurs in spring as well as in late summer to fall. Females bear from 2 to 18 young, usually in late summer. The juveniles at birth are about 225 mm (8.75 in.) long.

Cottonmouth
Agkistrodon piscivorus

760 to 1,800 mm (30 to 71 in.) This venomous snake, often confused with several large nonvenomous water snakes, has elliptical pupils, a prominent pit on each side of the snout, and mostly undivided subcaudals on the anterior portion of the tail. Dorsal ground color is usually olive or brown, and dark crossbands with light centers constitute the body pattern. Juveniles have bright chestnut markings and yellowish tail tips, but some large adults are dark and virtually patternless. The dorsal scales are keeled. Males often attain larger sizes than females.

Cottonmouths live in a variety of aquatic and semiaquatic habitats in southeastern Virginia and eastern North Carolina. They occur in much of South Carolina but are generally scarce outside the Coastal Plain.

Cottonmouths are active both night and day. When disturbed, some individuals quickly retreat; others coil, vibrate the tail, and gape the mouth widely. If further threat-

ened, most will bite, but they are not as aggressive as commonly believed. Bites are potentially serious and should receive prompt medical attention.

An opportunistic feeder, this snake eats fish, amphibians, reptiles, birds, and small mammals. Courtship behavior has been observed in spring and in late summer. The 5 to 11 young per litter average about 260 mm (10.25 in.) long and are usually born in late summer. Females may produce litters only every second or third year.

Eastern Diamondback Rattlesnake *Crotalus adamanteus*

1,067 to 1,980 mm (42 to 78 in.)
This huge, heavy-bodied rattlesnake has dark, diamond-shaped body blotches with yellowish margins and light centers. A dark brown to black bar, bordered above and below by a yellow line, extends from the eye to the mouth. Dorsal ground color varies from olive to dark brown. The dorsal scales are strongly keeled. Males often attain larger sizes than females.

The Diamondback inhabits the lower Coastal Plain of South Caro-

lina and the southeastern corner of North Carolina. Favorite habitats are pine flatwoods, brushy fields bordered by forests, and drier pocosins. These big rattlers frequently hide in stump holes, under brush piles, and in burrows of other animals. They are chiefly diurnal.

The Eastern Diamondback is an impressive animal. When aroused, it quickly assumes a defensive posture: body coiled, rattle erect and buzzing, neck flexed with head directly facing the source of annoyance. Though rare and infrequently encountered, it is potentially the most dangerous snake in the Southeast. Adults feed largely on rabbits, but rats, mice, and other small mammals are also eaten. The 7 to 29 young per litter, born usually in summer or early fall, are about 400 mm (15.75 in.) long. Females probably produce litters only every second or third year.

The Eastern Diamondback has suffered greatly from habitat loss and individual persecution. It is protected as an Endangered species in North Carolina.

Timber Rattlesnake (Coastal Plain and Piedmont "canebrake" form)

Timber Rattlesnake
Crotalus horridus

914 to 1,829 mm (36 to 72 in.) These are large, heavy-bodied rattlesnakes with dark blotches and wavy cross-bands. Timber Rattlers from the Mountains of the Carolinas and Virginia have a yellowish or black dorsal ground color and an average of 23 rows of scales around the middle of the body. Those in southeastern Virginia and most of the Carolinas are known as Canebrake Rattlesnakes. They typically have a brown, gray, or pinkish dorsal ground color with a reddish or brown middorsal stripe, an orange to dark brown bar from the eye to the rear of the mouth, and 25 rows of scales around the middle of the body. The largest individuals are males.

Timber Rattlesnakes once occurred throughout the area up to about 2,000 m (6,560 ft.) elevation, but they have been eliminated from much of the region due to extensive deforestation, agriculture, and urbanization. Favorite habitats are rocky hillsides, fields bordered by forests, river valleys and swamps, low pinewoods, and pocosins. In the Mountains, these rattlers aggregate during the fall and spend the winter in deep crevices of rock outcrops.

Individuals often hide in stump holes and under various cover on the surface, but they may be active

Timber Rattlesnake (high-elevation dark phase)

by day or night. Nocturnal activity is especially common during the hot summer months. When discovered, these rattlers usually remain still or attempt to escape. They will, however, vigorously defend themselves when provoked. Bites are occasionally fatal to humans and should receive prompt medical attention. Small mammals, especially rodents, constitute the primary food; some birds are also eaten. Mating has been observed in mid-August but may also take place in spring. The young, usually born in August and September, number from 4 to 20 per litter and are about 350 mm (13.75 in.) long. Most females probably give birth only every second or third year.

The Timber Rattlesnake has suffered greatly from habitat loss and deliberate persecution. It is listed as a species of Special Concern in North Carolina. Virginia's remaining Coastal Plain populations are state-listed as Endangered.

Pigmy Rattlesnake
Sistrurus miliarius

380 to 660 mm (15 to 26 in.) These small, moderately slender-bodied rattlesnakes have large scales on top of the head, a conspicuous pit between the eye and nostril, a slender tail, and a tiny rattle that produces a faint sound similar to the buzz of some insects. Dorsal color varies from gray to red, with prominent dark brown or black blotches, and many specimens have a reddish middorsal stripe. The venter is white to pink with brown or black markings. The dorsal scales are keeled. Juveniles have a white or yellow tail tip and generally brighter patterns than adults. In extreme southeastern South Carolina, individuals are larger and darker than those from other portions of our area. Pigmy Rattlesnakes living on and near the Albemarle Peninsula in North Carolina have bold, glossy patterns of red, orange, or pink. The largest individuals are usually males.

This species occurs in most of South Carolina and in southern

and eastern North Carolina to Hyde and Tyrrell counties. Pine flatwoods and sandy, open woodlands with pines, Wiregrass, and scrub oaks are preferred habitats. In these places, Pigmy Rattlesnakes frequently live around cypress ponds and other bodies of water. Individuals may be active on the surface both day and night, but they are often found beneath logs and other surface cover.

When disturbed, a Pigmy Rattlesnake usually attempts to crawl away, but it will usually strike quickly if aggravated. This species is the smallest and, along with the Copperhead, the least dangerous of our venomous snakes. Nevertheless, its bite frequently produces secondary infection and should be given prompt medical care. Frogs, lizards, and small snakes and mammals are the primary prey. The 3 to 9 young per litter are about 170 mm (6.75 in.) long and are born usually in August and September.

The Pigmy Rattlesnake is protected as a species of Special Concern in North Carolina.

Glossary

Adaptation Any trait (anatomical, physiological, or behavioral) that makes an organism more fit to survive and reproduce in a particular habitat. Also, the process leading to the formation of such a trait.

Amphipods Small crustaceans with compressed bodies. They live in and on bottom debris.

Anterior The front end of an animal.

Arthropods Animals with segmented bodies and jointed appendages; the insects, crustaceans, and their relatives.

Axillary Pertaining to the axilla, armpit.

Barbel Small, fleshy extension on the chin or throat, as in some turtles.

Binomial Nomenclature The system, first devised by Linnaeus, of assigning every organism a two-part Latinized name—i.e., genus and species.

Biodiversity The variety of life forms within a region.

Biogeography The study of the geographical distributions of living organisms.

Bridge Part of the plastron that joins onto the carapace in turtles.

Bufonids The common toads with dry, warty skin and two digging tubercles on each hind foot.

Caiman Genus containing five species of crocodilians native to Central and tropical South America. *Caiman crocodilus*, the Spectacled Caiman, has been widely introduced into the United States.

Carapace The upper part of the turtle shell, including its bones and horny scutes.

Caudal Pertaining to the tail, situated toward the hind part of the body.

Cenozoic Era The last 63 million years of the earth's history; mammals, birds, and flowering plants dominate.

Chemoreceptors Smell and taste receptors (sensory cells or organs).

Cirrus A small, fingerlike projection on the upper lip of males of some plethodontid salamanders.

Cloaca The chamber into which the large intestine and the urogenital ducts empty their contents.

Communal Pertaining to the members of a group that may cooperate in building or in using a nest but not in subsequent care of the eggs or young.

Conspecific Belonging to the same species.

Cornification A process wherein cells in the epidermis become thin, flat, and horny. They form scales or a thin, nonliving layer that is periodically shed.

Costal Grooves The vertical furrows on the sides of most salamanders.

Cranial Crest Ridge on the head of a toad (*Bufo*).

Cretaceous Last geological period of the Mesozoic era, lasting about 75 million years. Dinosaurs reached their peak and became extinct.

Crossband A broad area of color

oriented with its long axis perpendicular to the body axis.

Crossbar A marking (bar or stripe) that goes across rather than down the body.

Crustaceans A large group of the arthropods including crayfish and their relatives.

Cryptic Hidden, concealed, unrecognized.

Cusp Pointed projection on a tooth or the jaw's edge.

Dermis The inner layer of the skin. It is fibrous, vascular, and sensitive.

Devonian A geological period of the Paleozoic era. It occurred about 370 million years ago and is referred to as the Age of Fishes.

Digital Pad Enlarged, disklike structure at the tips of the digits of many hylid frogs.

Diploid Having two copies of each chromosome (one from each parent).

Disjunct Discontinuous, separated, not contiguous; usually refers to spatially isolated populations that formerly were parts of a large, wide-ranging population.

Dorsolateral The region along the body where the side joins the back.

Ecology The study of the interaction of organisms with the physical environment and the other organisms that live there.

Eft Brightly colored, terrestrial, subadult stage of a newt.

Embryology Study of early development of organisms: their formation and growth.

Endemic A species or other taxon that is native to a particular place and found nowhere else.

Fall Line Boundary between the older and harder rocks of the Piedmont and the loose sediments of the Coastal Plain. It is marked by falls or rapids along streams flowing into the Coastal Plain.

Fauna The animals of a region.

Flatwoods Coastal Plain habitat dominated by pine (mainly Longleaf or Loblolly) and Wiregrass. Relatively open woodlands.

Fossorial Adapted for digging: in salamanders and lizards, the long bodies and small limbs; in toads, the digging tubercle (spade) on each hind foot; in snakes, the smooth scales and modified snout.

Genetic Drift Evolution by chance rather than by selection; the tendency within small populations for heterozygous gene pairs to become homozygous, thus reducing genetic variability.

Gular Relating to the throat.

Habitat The organisms and physical environment in a particular place.

Hedonic Stimulating animals to mate; erotic.

Herpetology The scientific study of amphibians and reptiles.

Humeral Scutes Anteriormost pair of scales on the plastron; each is located between the gular and a pectoral scale.

Hybridize To produce offspring by mating of individuals from different species.

Hybrid Swarm The many hybrid individuals resulting from a temporary or local breakdown of reproductive isolation between related species.

Hylids The common treefrogs of our

area. This group also includes the cricket and chorus frogs.

Imbricate To overlap in a regular order.

Keel A prominent longitudinal ridge.

Kilogram (kg) Equal to 2.2 lbs.

Kilometer (km) Equal to 0.62 miles.

Labial Pertaining to the lip.

Larva An immature stage that differs markedly from the adult.

Life Cycle The entire life span of an individual from origin to reproduction.

Life History The series of stages and activities through which an organism normally passes from zygote formation to death.

Marginal Scutes Scales located along the periphery of the carapace.

Mental Gland A roundish to ovoid bulge on the anterior chin of certain male salamanders. Its secretion facilitates courtship.

Mesic Moderately moist.

Mesozoic Era This important chapter in the history of the earth began 230 million years ago and lasted 165 million years; reptiles and gymnosperms dominated.

Metamorphosis A marked change in appearance, as when a tadpole becomes a frog. This transformation is less conspicuous in salamanders.

Microhylids The narrow-mouthed toads with smooth, moist skin and pointed head with skin fold; small and secretive.

Mimic To imitate in form, color, or behavior. Usually an edible species that takes on characteristics of a species noxious to predators.

Mollusks Members of the phylum Mollusca, including the snails and clams.

Molt The casting off (shedding) of the outer layer of epidermis.

Morphology The form and structure of an organism.

Nasolabial Groove A tiny furrow from nostril to lip of plethodontid salamanders. It facilitates olfaction.

Neonate Newly hatched, newborn.

Newt A salamander in the family Salamandridae; the adults lack conspicuous costal grooves and gills but have lungs.

Nictitating Membrane The third eyelid, a thin membrane that is often transparent and capable of being drawn from the inner angle over the outer surface of the eye.

Nuchal Related to the back of the neck.

Nuchal Scute The medial scale of the carapace located near the neck and anterior to the vertebrals, sometimes referred to as the cervical or precentral scute.

Nucleolus A roundish, dense structure in the nucleus. It is rich in ribonucleic acid (pl. nucleoli).

Ocellus Eyelike spot of color (pl. ocelli).

Osteoderms Bones in the dermis of some reptiles; usually found beneath scales, where they serve as added armor.

Oviparous Pertaining to species that lay eggs.

Paedomorphosis The retention of embryonic or larval features by adults, an evolutionary process under genetic control.

Paleozoic Era An early major subdivision of geological time characterized by an abundance of macrofossils: invertebrates, fish, amphibians, clubmosses, horsetails, and ferns. It extended from 600 to 230 million years ago.

Parotoid Large, wartlike gland near the tympanum of toads.

Parthenogenesis The production of an organism from an unfertilized egg; unisexual reproduction.

Pectoral Scutes Paired scales of plastron, located just behind the humeral scales.

Pelagic Living in or pertaining to the open waters of the sea.

Pelobatids The spadefoot toads with smooth, moist skin and only one digging tubercle on the hind foot; very secretive.

Phenotype The visible traits developed under the influence of an organism's genes and its environment.

Pheromone A glandular secretion that evokes specific behavior in another individual after tasting or smelling the secretion.

Phylogenetic Pertaining to evolutionary relationships and lines of descent; the origin and evolution of higher taxa.

Physiographic Pertaining to physiography—the study of the earth's surface (relief).

Physiology The study of the internal functions and processes of organisms and their parts.

Plastron The bones and horny scutes that form the ventral part of a turtle's shell.

Plethodontids Salamanders in the family Plethodontidae. They lack lungs, have a unique nasolabial groove, and are common in the Western Hemisphere.

Pleural Referring to lungs or to chest wall.

Pleural Scutes Large scales on the carapace of turtles, located between vertebral and marginal scutes; referred to as costals by some authors.

Pocosin An upland swamp in the Coastal Plain, especially an evergreen shrub bog.

Population A group of conspecific individuals occupying a particular space at the same time.

Posterior The rear of an animal.

Postocular Located posterior to the eye.

Prefrontal Large, usually paired dorsal scales located on snout between frontal and internasals.

Primitive Referring to traits that evolved early and later gave rise to other traits; ancestral. Primitive traits are usually, but not always, less complex.

Respire To breathe, to exchange gases (as between an organism or a cell and its environment).

Reticulate Having a netlike pattern.

Rugose Having fine wrinkles or raised longitudinal lines.

Savanna A plain with scattered trees and drought-resistant undergrowth, mostly grasses.

Scale Rows The longitudinal rows of scales around the bodies of lizards and snakes. They are counted just anterior to the middle of the body. In snakes, the count starts with the row adjacent to the ventrals and continues diagonally over the back to the ventrals on the opposite side.

Scute A large scale.

Serrate Notched, sawlike.

Sibling Species Two or more closely related species, morphologically similar but reproductively isolated.

Smooth Scale A scale without a longitudinal ridge (keel).

Spartina Widely distributed genus of grasses, common in salt marshes along the coast.

Speciation The process wherein a population becomes genetically diverse and new species are formed.

Species The basic unit of taxonomy. A population or a group of populations of closely related and similar organisms that are capable of interbreeding.

Spermatophore A small, gelatinous mass that bears a packet of sperm on top; it is produced by many species of salamanders.

Sphagnum Large genus of mosses that grow in wet, acid areas.

Subadult An individual that is not yet sexually mature; e.g., a transformed amphibian.

Subcaudal Located on the underside of the tail.

Subspecies A race or a subdivision of a species that is geographically distinct.

Sympatric Referring to those populations whose geographic ranges overlap.

Talus Rock debris or fragments at the base of a slope or cliff.

Taxon Any taxonomic group, e.g., order, family (pl. taxa).

Taxonomy The science of naming, describing, and classifying organisms into categories: phylum, class, order, family, genus, species. It is based on established principles and procedures.

Territoriality Defense of an area against the entry of other members of the same species; usually involves males.

Territory A defended area.

Tetraploid Having four complete chromosome sets (four times the haploid number of chromosomes) in the cell nucleus.

Tetrapods Vertebrates that typically have two pairs of limbs: the amphibians, reptiles, birds, and mammals.

Thecodontia Order of primitive, Triassic reptiles that gave rise to dinosaurs, pterodactyls, crocodilians, and birds.

Transform To undergo metamorphosis, to change from larval to subadult stage.

Trigonal Triangular.

Trill A frog call in which the same note is rapidly repeated (8–30 times per second).

Troglobite An animal modified for cave or subterranean life, usually lacking pigment and functional eyes.

Truncate Lacking a point, squarish.

Tubercle A small, knoblike projection or wart.

Tympanum Eardrum.

Vent The cloacal aperture, the posterior body opening.

Ventrals Large, transverse scales of snakes, useful in locomotion.

Vertebral Scutes The large medial scutes of the carapace on turtles. They overlay the bony shell below.

Vocal Sac Inflatable, elastic pouch on or near the chin of male frogs and toads. A resonating chamber.

Wrack Dried seaweed or other debris above the tide line on beaches.

Xeric Arid, lacking moisture.

Useful References

Altig, Ronald. 1970. A key to the tadpoles of the Continental United States and Canada. *Herpetologica* 26:180–207.

Altig, Ronald, R. W. McDiarmid, K. A. Nichols, and P. C. Ustach. 1998. A key to the anuran tadpoles of the United States and Canada. *Contemporary Herpetology Information Series*, <http://www.contemporaryherpetology.org/chis/>.

Bailey, Mark A., Jeffrey N. Holmes, Kurt A. Buhlmann, and Joseph C. Mitchell. 2006. *Habitat Management Guidelines for Amphibians and Reptiles of the Southeastern United States.* Technical Publication HMG-2, Partners in Amphibian and Reptile Conservation, Montgomery, Ala.

Bartlett, Richard D., and Patricia P. Bartlett. 2006. *Guide and Reference to the Amphibians of Eastern and Central North America (North of Mexico).* University of Florida Press, Gainesville.

Behler, John L., and F. Wayne King. 1985. *The Audubon Society Field Guide to North American Reptiles and Amphibians.* Knopf, New York.

Bishop, Sherman C. 1947. *Handbook of Salamanders.* Comstock Publishing Co., Ithaca, N.Y.

Braswell, Alvin L., William M. Palmer, and Jeffrey C. Beane. 2003. *Venomous Snakes of North Carolina.* North Carolina State Museum of Natural Sciences, Raleigh, N.C.

Buhlmann, Kurt, Tracey Tuberville, and Whit Gibbons. 2008. *Turtles of the Southeast.* University of Georgia Press, Athens.

Carr, Archie F. 1963. *The Reptiles.* Time, New York, N.Y.

Catalogue of American Amphibians and Reptiles. Published by the Society for the Study of Amphibians and Reptiles, <www.herplit.com/SSAR>.

Catesbeiana. Published biannually by the Virginia Herpetological Society, <http://fwie.fw.vt.edu/VHS/index.html)>.

Conant, Roger, and Joseph T. Collins. 1998. *A Field Guide to Reptiles and Amphibians of Eastern and Central North America.* 3rd ed., expanded. Houghton Mifflin, Boston.

Copeia. Published quarterly by the American Society of Ichthyologists and Herpetologists, <www.asih.org>.

Crother, Brian I., ed. 2008. *Scientific and Standard English Names of Amphibians and Reptiles of North America North of Mexico.* 6th ed. Society for the Study of Amphibians and Reptiles Circular 37.

Dodd, C. Kenneth. 2004. *The Amphibians of Great Smoky Mountains National Park.* University of Tennessee Press, Knoxville.

Dorcas, Michael E. 2004. *A Guide to the Snakes of North Carolina.* Davidson College, Davidson, N.C.

Dorcas, Michael E., Steven J. Price, Jeffrey C. Beane, and Sarah Cross Owen. 2007. *The Frogs and Toads* of North Carolina: Field Guide and Recorded Calls. North Carolina Wildlife Resources Commission, Raleigh.

Dorcas, Mike, and Whit Gibbons. 2008. Frogs and Toads of the Southeast. University of Georgia Press, Athens.

Ernst, Carl H. 1992. Venomous Reptiles of North America. Smithsonian Institution Press, Washington, D.C.

Ernst, Carl H., Jeffrey E. Lovich, and Roger W. Barbour. 1994. Turtles of the United States and Canada. Smithsonian Institution Press, Washington, D.C.

Gans, Carl, and T. S. Parsons, eds. 1969–2008. Biology of the Reptilia. Vols. 1–20. Academic Press, New York, N.Y., and the Society for the Study of Amphibians and Reptiles.

Gibbons, J. Whitfield, and Michael E. Dorcas. 2004. North American Watersnakes: A Natural History. University of Oklahoma Press, Norman.

Gibbons, Whit. 1983. Their Blood Runs Cold. University of Alabama Press, Tuscaloosa.

Gibbons, Whit, and Mike Dorcas. 2005. Snakes of the Southeast. University of Georgia Press, Athens.

Gibbons, Whit, and Patricia J. West, eds. 1998. Snakes of Georgia and South Carolina. Savannah River Ecology Laboratory Herp Outreach Publication #1, Aiken, S.C.

Herpetologica. Published quarterly by the Herpetologists' League, <www.herpetologistsleague.org>.

Herpetological Conservation and Biology, <http://www.herpconbio .org/>.

Herpetological Review. Published quarterly by the Society for the Study of Amphibians and Reptiles, <www.herplit.com/ SSAR>.

Huheey, James E., and Arthur Stupka. 1967. Amphibians and Reptiles of Great Smoky Mountains National Park. University of Tennessee Press, Knoxville.

Journal of Herpetology. Published quarterly by the Society for the Study of Amphibians and Reptiles, <www.herplit.com/ SSAR>.

Klauber, Laurence M. 1972. Rattlesnakes: Their Habits, Life Histories, and Influence on Mankind. 2nd ed. Vols. 1 and 2. University of California Press, Berkeley.

Lannoo, Michael J., ed. 2005. Amphibian Declines: The Conservation Status of United States Species. University of California Press, Berkeley.

Lillywhite, Harvey B. 2008. Dictionary of Herpetology. Krieger Publishing Co., Malabar, Fla.

Martof, Bernard S., William M. Palmer, Joseph R. Bailey, and Julian R. Harrison III. 1980. Amphibians and Reptiles of the Carolinas and Virginia. University of North Carolina Press, Chapel Hill.

Minton, Sherman A., Jr., and Madge R. Minton. 1969. Venomous Reptiles. Charles Scribner's Sons, New York.

Mitchell, J. C. 1994. The Reptiles of Virginia. Smithsonian Institution Press, Washington, D.C.

Mitchell, Joe, and Whit Gibbons. In press. *Salamanders of the Southeast*. University of Georgia Press, Athens.

Mitchell, Joseph C., Alvin R. Breisch, and Kurt A. Buhlmann. 2006. *Habitat Management Guidelines for Amphibians and Reptiles of the Northeastern United States*. Technical Publication HMG-3, Partners in Amphibian and Reptile Conservation, Montgomery, Ala.

NC Herps, the North Carolina Herpetological Society Newsletter. Published quarterly by the North Carolina Herpetological Society, <www.ncherps.org>, North Carolina State Museum of Natural Sciences, 11 West Jones St., Raleigh, N.C. 27601-1029.

Palmer, William M., and Alvin L. Braswell. 1995. *Reptiles of North Carolina*. University of North Carolina Press, Chapel Hill.

Parker, H. W., and A. Grandison. 1977. *Snakes: A Natural History*. British Museum, London.

Petranka, James W. 1998. *Salamanders of the United States and Canada*. Smithsonian Institution Press, Washington, D.C.

Pough, F. Harvey, Robin M. Andrews, John E. Cadle, Martha L. Crump, Alan H. Savitsky, and Kentwood D. Wells. 2004. *Herpetology*. 3rd ed. Prentice Hall, Upper Saddle River, N.J.

Pritchard, Peter C. H. 1967. *Living Turtles of the World*. T. F. H. Publications, Jersey City, N.J.

Rossi, John V., and Roxanne Rossi. 2003. *Snakes of the United States and Canada: Natural History and Care in Captivity*. Kreiger Publishing Co., Malabar, Fla.

Stebbins, Robert C., and Nathan W. Cohen. 1995. *A Natural History of Amphibians*. Princeton University Press, Princeton, N.J.

Wells, Kentwood D. 2007. *Ecology and Behavior of Amphibians*. University of Chicago Press, Chicago, Ill.

Wright, Albert H., and Anna A. Wright. 1949. *Handbook of Frogs and Toads of the United States and Canada*. 3rd ed. Comstock Publishing Co., Ithaca, N.Y.

Wright, Albert H., and Anna A. Wright. 1957. *Handbook of Snakes of the United States and Canada*. Vols. 1 and 2. Comstock Publishing Co., Ithaca, N.Y.

Zim, Herbert S., and Hobart M. Smith. 2001. *Reptiles and Amphibians: A Golden Guide*. St. Martin's Press, New York, N.Y.

Zug, George R., and Carl H. Ernst. 2004. *Smithsonian Answer Book, Snakes*. Smithsonian Institution Press, Washington, D.C.

Zug, George R., Laurie J. Vitt, and Janalee P. Caldwell. 2001. *Herpetology: An Introductory Biology of Amphibians and Reptiles*. 2nd ed. Academic Press, San Diego, Calif.

Additional Websites of Interest

The Charleston Museum: <http://www.charlestonmuseum.org/topic.asp?id=1>.

Davidson College: Amphibians and Reptiles of North Carolina, <http://www.bio.davidson.edu/projects/herpcons/herpcons.html>; Carolina Herp Atlas,

<http://www.carolinaherpatlas
.org/>.
North Carolina Herpetological
Society: <http://www.ncherps
.org/>.
North Carolina Partners in
Amphibian and Reptile
Conservation: <http://www
.ncparc.org/>.
North Carolina State Museum of
Natural Sciences: <http://www
.naturalsciences.org/>.
Partners in Amphibian and Reptile
Conservation: <http://www
.parcplace.org/>.

Savannah River Ecology Laboratory:
<http://www.uga.edu/~srel/>.
South Carolina State Museum:
<http://www.museum.state
.sc.us/>.
Southeast Partners in Amphibian
and Reptile Conservation:
<http://www.separc.org/>.
Virginia Herpetological Society:
<http://fwie.fw.vt.edu/VHS/index
.html>.
Virginia Museum of Natural History:
<http://www.vmnh.net/>.
Virginia Natural History Society:
<http://fwie.fw.vt.edu/VNHS/>.

Photo Credits

All photographs are by Jack Dermid except for the following:

Jeffrey Beane: Cumberland Slider (profile), Southeastern Five-lined Skink, Eastern Hognose Snake (feigning death)

Alvin Braswell: Marbled Salamander, Northern Dusky Salamander, Shovel-nosed Salamander, Chamberlain's Dwarf Salamander, Blue Ridge Two-lined Salamander, "Sandhills Eurycea", Tellico Salamander, White-spotted Slimy Salamander, Cope's Gray Treefrog, Squirrel Treefrog (brown/spotted phase), Green Frog, Southern Leopard Frog, Striped Mud Turtle, Painted Turtle, River Cooter, Red-eared Slider, Spiny Softshell, Mediterranean Gecko, Green Anole (green phase), Five-lined Skink (young adult), Mimic Glass Lizard, Eastern Glass Lizard (adult male), Ring-necked Snake (Northern), Mole Kingsnake, Red-bellied Water Snake (adult), Banded Water Snake, Northern Water Snake, Carolina Water Snake, Brown Water Snake, Copperhead (adult), Copperhead (juvenile), Cottonmouth, Cottonmouth (close-up, gaping), Timber Rattlesnake ("canebrake" phase)

Matthew Godfrey: Leatherback Sea Turtle

Jeff Hall: Eastern Diamondback Rattlesnake

Robert Palmatier: Kemp's Ridley Sea Turtle

Peter Richardson: Hawksbill Sea Turtle

Nathan Shepard: Bog Turtle

Wayne Van Devender: Green Salamander, Allegheny Mountain Dusky Salamander, Blue Ridge Dusky Salamander, Spotted Dusky Salamander, Dwarf Black-bellied Salamander, Virginia Dusky Salamander, Santeetlah Dusky Salamander, Northern Two-lined Salamander, Dwarf Salamander, Blue Ridge Gray-cheeked Salamander, Cheoah Bald Salamander, South Mountain Gray-cheeked Salamander, Southern Gray-cheeked Salamander ("Clemson" phase), Northern Gray-cheeked Salamander, Chattahoochee Slimy Salamander, Northern Slimy Salamander, South Carolina Slimy Salamander, Cumberland Plateau Salamander, Big Levels Salamander, Southern Appalachian Salamander, Shenandoah Mountain Salamander, Gray Treefrog, New Jersey Chorus Frog, Pickerel Frog

Index

Manteo Library
East Albemarle Regional Library
Manteo, NC 27954
252-473-2372

SEP -- 2011